Studies in Applied Philosophy, Epistemology and Rational Ethics

Volume 18

About this Series

Studies in Applied Philosophy, Epistemology and Rational Ethics (SAPERE) publishes new developments and advances in all the fields of philosophy, epistemology, and ethics, bringing them together with a cluster of scientific disciplines and technological outcomes: from computer science to life sciences, from economics, law, and education to engineering, logic, and mathematics, from medicine to physics, human sciences, and politics. It aims at covering all the challenging philosophical and ethical themes of contemporary society, making them appropriately applicable to contemporary theoretical, methodological, and practical problems, impasses, controversies, and conflicts. The series includes monographs, lecture notes, selected contributions from specialized conferences and workshops as well as selected Ph.D. theses.

More information about this series at http://www.springer.com/series/10087

Pentti Määttänen

Mind in Action

Experience and Embodied Cognition in Pragmatism

Springer

Pentti Määttänen
Department of Philosophy, History,
Culture and Art Studies
University of Helsinki
Helsinki
Finland

ISSN 2192-6255 ISSN 2192-6263 (electronic)
Studies in Applied Philosophy, Epistemology and Rational Ethics
ISBN 978-3-319-17622-2 ISBN 978-3-319-17623-9 (eBook)
DOI 10.1007/978-3-319-17623-9

Library of Congress Control Number: 2015936622

Springer Cham Heidelberg New York Dordrecht London

Printed on acid-free paper

Springer International Publishing AG Switzerland is part of Springer Science+Business Media
(www.springer.com)

Contents

Introduction

Challenging Classical Dichotomies

Pragmatism was born as a challenge to the presumptions of classical philosophy. Charles Peirce presented his pragmatic maxim as an alternative to René Descartes' thought that clear and distinct ideas can be found by using introspection. John Dewey's mature work is devoted to the reconstruction of philosophy. The main idea of pragmatism is that experience as sense perception is a too narrow view. Action must be included in the concept of experience. This requires a different view about the structure of experience and the object of knowledge. The earlier view, according to which experience is sense perception and the object of knowledge consists of the hidden causes of perceptions, is not adequate if the role of action in experience is to be taken into account. In pragmatism to know is to know what to do, and the object of knowledge consists of the possibilities to act in given circumstances. The main objective of this book is to find out what would follow if this revision were consistently carried out.

Classical presumptions, which were formed during the history of philosophy, are still prevailing in the background of contemporary discussion. The dichotomy of apparent and real and the dichotomy of internal and external belong to the most influential ones. These presumptions are not necessarily valid. The dichotomy of apparent and real emerged in ancient Greece when pre-Socratic philosophers had difficulties in conceptualising movement and change. The puzzle with Achilles and the tortoise seemed to be a serious one, which made movement hard to understand. The idea of genuine change seemed to be in danger because the basic realities are what they are and cannot be otherwise, or so it was claimed. Some philosophers became convinced that the real world ultimately consists of unchanging entities, and this led to the conviction that genuine knowledge is about these fixed and immutable entities.

On the other hand, nature is full of movement and change, as everyone can perceive. Therefore the experienced world cannot be properly conceived and thus is something "less real", only apparent. The doctrine of ideas presented by Socrates and Plato consolidated the dichotomy of the empirical world as something apparent and the world of fixed and immutable ideas as something real. Genuine knowledge concerns the ideas. True knowledge about ideas can be attained by rational discussion, not by observing what happens in nature. It is important to note that the real, as contrasted to the apparent, is in principle, by definition, beyond the scope of our epistemic access to the empirical world.

The original motivation for contrasting apparent and real is gone. Insufficient understanding of mathematics caused the puzzle with Achilles, and movement and change has been conceptualised successfully in natural science. Therefore it is reasonable to re-evaluate the dichotomy of apparent and real.

The dichotomy of external and internal was established by René Descartes, John Locke and their followers. Internal consciousness perceives the so-called external world through sense organs. The outcome of this approach is a conception of the structure of experience and the object of knowledge that is still dominating philosophical disputes. External world is the cause of sense perceptions, which are connected to ideas as internal units. Knowledge is contained in ideas (internal symbols, representations or other) and their combinations. The ultimate object of knowledge is the hidden causes of perceptions. They are hidden because, according to this classical view of the structure of experience, all we can perceive are the perceptions, the effects of the external causes. Effects are not similar to the causes; neither are there other ways of inferring the character of the causes from the character of the effects. However, true knowledge is supposed to be about the mind-independent real world beyond our epistemic access through the senses. A great deal of contemporary views still holds this dichotomy. The brain as an organ of thought is the seat of mental activities. Perceptions, experiences, qualia and so on are commonly conceived to reside in the brain.

Philosophical naturalism is one of the topics where the rejection of the dichotomies mentioned above leads to alternative viewpoints, which are discussed in Chap. 1. Most contemporary naturalists maintain the classical view concerning the character of experience. Natural science is assumed to give a theory about the ultimate constituents of the real world as opposed to everyday experience. The human mind tries to form a conception of this external reality. This is what John Dewey called the spectator theory of knowledge. However, this view of experience does not follow from the stand that human beings have evolved within nature, which is one way to characterize naturalism. A clear alternative to hard naturalism emphasizing the role of natural science is soft naturalism, which does not put nature and culture against each other. Culture is a product of nature, and human beings are natural creatures within nature. There is an ontological symmetry between human bodies and all other elements in nature; they are all made of the same stuff.

Naturalism entails a critical attitude towards classical epistemology as *a priori* conceptual analysis, but it seems to be a kind of *a priori* decision to commit oneself to natural scientific methods in all knowledge. These methods have been developed during centuries for solving problems typical for physics, chemistry and so on, but the problems in gaining knowledge about human culture and society are different. Instead of commitment to natural scientific methods one can take science as a problem solving enterprise. Any method can be used if it can be assumed to give information that is relevant for solving the problem in question.

This view of science as problem solving fits well with the revised notion of the structure of experience, which is discussed in Chap. 2. The classical view maintains that experience is based on perception. The real world is a swarm of hidden causes of perceptions, on the ground of which knowledge is attained. The world is perceived as individual objects, their properties and mutual relations. The pragmatist view of experience is different. Human beings are embodied creatures, which are in constant interaction with other elements in the world. The world is experienced as possibilities of action. The hidden causes of perception are not the object of knowledge. The structure of experience and the proper object of knowledge can be expressed with the simple scheme: $\mathbf{S_1} \rightarrow \mathrm{O} \rightarrow \mathbf{S_2}$. The situation $\mathbf{S_1}$ is problematic, and some operations O have to be performed in order to attain the situation $\mathbf{S_2}$. The latter situation is hidden at first, but not in principle. The object of knowledge is the relation between these situations, and this relation is mediated by controlled activity, certain operations. To know is to know what to do in the situations one encounters in the world. This object of knowledge is within our epistemic access, and there is no need to contrast it with something "more real" beyond the scope of our epistemic access.

If knowing is about what to do, then the next question concerns inevitably the vehicles of knowing, the vehicles of cognition. The classical view and its followers appeal to ideas, mental contents, internal symbols or representations, propositions and so on. The pragmatist alternative to this tradition is, according to Charles Peirce, the notion of habit of action, which is the topic of Chap. 3. Habits are beliefs, but they are not internal units, properties of the brain or the body. Habits are modes of interaction, structured schemes of action, which are formed when action accommodates to objective conditions of action. Habits become to be beliefs about those conditions of action.

Beliefs are supposed to be general entities, and habits of action fulfil this condition as being repeatable indefinitely many times in the future. The mode of existence of habits is somewhat peculiar because the essential feature of habits is just this, *esse in futuro*, as Peirce put it. This entails that habits can only be objects of thought. However, nothing like Cartesian dualism follows because all instances of habits, all performed acts, take place in this material world we inhabit. In this view, cognition is basically anticipation of action.

Chapter 4 deals with habits of action as meanings. Charles Peirce's definition, according to which what a thing means is simply what habits it involves, can be applied to any object of perception. This entails that the rather common comparison between words and tools is more than a mere analogy. Everyday objects as chairs, hammers and so on are meaningful entities of their own right. A habit of action associated with these kinds of things enables one to anticipate what consequences will probably follow if one applies the habit in question. This gives cognitive distance to the prevailing situation, which is a minimum requirement of anything's being a meaningful entity. This notion of tacit (non-linguistic) meaning gives grounds to argue for a system of meanings quite independently of any linguistic competence. These habits are thus vehicles of meaningful cognition.

Linguistic meaning can be explained by the same principle. Ludwig Wittgenstein's definition that the meaning of a linguistic expression is the way it is used in a linguistic community is a special case of Peirce's definition since the use of an expression is surely a habit of action. The system of meanings thus consists of two kinds of meanings: non-linguistic tacit meanings and linguistic meanings, which are formed on the basis of tacit meanings. Of course, the ability to use natural language has an enormous effect on the system of meanings as a whole. The main point is, however, that language is not the only vehicle of meaningful cognition. The bottom-up analysis of the system of meanings is one way to give a naturalistic explanation of the emergence of language. After all, natural language is natural, not supernatural.

The layered system of meanings gives a new perspective to the traditional conception according to which, quite correctly, experience is always interpreted with meanings. The Kantian tradition emphasizes that meanings are internal entities. Concepts carve up the nature. However, tacit meanings are not so conventional as linguistic meanings. Tacit meanings are formed when action accommodates to objective conditions of action. David Hume already wondered whether sceptics, while being extremely sceptical about the existence of the external world, would use a door or a window when exiting a room (Hume 2001, p. 20). Doors and windows have different meanings, and this is based on conditions that are in no way conventional. This should be taken into account in developing a consistent system of meanings.

The rejection of the dichotomy of internal and external leads to an emphasis of organism environment interaction. Chapter 5 contains critical comments on the notion of internal representation and argues for the view that in searching for an alternative conception of mind and consciousness, the correct unit of analysis is the loop of action and perception. This is fully consistent with naturalism, since action and perception are realized through physical causal processes. However, the loop structure of habitual action gives habits the ability to anticipate future consequences of action. Habits are formed when similar action is repeated in similar circumstances, and when one faces these circumstances again, one can anticipate similar future experiences. In this way the (anticipated) future can have an effect on the present but not on the past, as Charles Peirce put it, meaning that no awkward backwards causation is required. This entails that habit of action is a teleological concept. This has the consequence that intentionality can be defined with the loop of perception and action, which gives a clear alternative to the attempts to attribute intentionality to internal brain states or processes.

In this view, mind is not a property of the brain or even the body. Mind is a property of organism environment interaction as characterized with the loop of perception and action. Any attempt to separate one element from this loop has the consequence that mentality is lost away. Drop the loop, lose the mental. And the relation between physical and mental causes of action turns out to be more complicated than in the attempts to look for causes of action in the brain.

The redefinition of the structure of experience leads to a redefinition of the notion of a matter of fact, and this has consequences concerning the relation between facts and values discussed in Chap. 6. Experience is about the possibilities of action, about what operations are required for getting from one situation to another. The object of knowledge is defined as a relation between two situations mediated by habitual action, and facts must be defined accordingly as relations between what is faced here and now and what will be faced in the future.

Contrary to David Hume, who could not literally perceive values in any separate situation and therefore separated sharply values from the world of facts, we can compare between possible courses of action and evaluate them on the ground of anticipated experiences as outcomes of these courses of action. Every experiential situation contains numerous possibilities of action, and it is impossible to execute all of them at the same time. One simply has to choose between these possibilities, and to choose is to value. Some anticipated consequences are valued higher or lower than the others. In other words, an acting agent is necessarily a valuing agent; facts and values are intertwined in experience.

Of course, not all choices are based on moral deliberation. The theory of value is constructed with a similar bottom-up strategy as the system of meanings. As natural creatures we have to do things simply in order to stay alive. These facts give certain values, and this valuation can be called biotechnical normativity. Higher values concerning humans as social and cultural beings are constructed on this basis. The outcome is a layered system of values, which cannot be analysed with any single moral rule or definition of moral good.

The last chapter deals with what follows from the previous considerations for the problems of truth and realism. Classical (or semantic) notion of truth as correspondence is a relation between meaningful statements and the world. Some versions of scientific realism use this notion in defining realism. The aim of science is to form theories about the mind-independent real world. Any attempt to talk about epistemic truth as a relation within our epistemic access to the world is labelled as antirealism because of the inability to access the mind-independent world. Epistemic access depends on mental capacities or internal conditions. Epistemic access is also tied to point of view, and this, in its turn, is a hindrance in aiming at objective knowledge. Perspective corrupts objectivity, or so it is said.

However, what mind-independence means depends on what one means with mind. If the classical view of mind as something internal is rejected and replaced by mind as a property of organism environment interaction, then mind-independence comes to mean independency of the scope of our interaction with the world as an epistemic access to the world. This interaction is not only sense perception. It reaches further than that. We interact with the theoretical objects of science with experimental devices. This interaction is epistemic in the sense that this is how we gain knowledge about the outcomes of interaction and also in the sense that the outcomes depend on the character of the physical instruments used. This can be called instrumental phenomenology.

There are two kinds of instruments used in our epistemic access to the world, bodily organs and external devices (external to the body, that is). They are as real and objective elements of the material world as any other element. They determine a physical viewpoint that must be distinguished from the conceptual viewpoint.

Mind is embodied, and this physical viewpoint cannot be avoided. Truth can be defined as correspondence between statements and the world. Everyday objects and the theoretical objects of science are within the reach of epistemic access, either by bodily organs or experimental devices. This can be called embodied epistemic truth. And there is an explanation of the fit between statements and the world: the fit is operational. There is no room for accusations of antirealism because the theoretical objects within the reach epistemic interaction are precisely the objects that experimental science deals with and there is no reason to deny the existence of things beyond the scope of present epistemic interaction. But they can become objects of knowledge only after they are reached with experimental devices, as the history of science shows.

Chapter 1
Philosophical Naturalism

The general philosophical framework of this book is philosophical naturalism. However, naturalism can be characterized in different ways that are not compatible with each other. It is, therefore, more than appropriate to specify what is here meant by naturalism.

Naturalism gets its basic content and motivation from the recognition of the fact that human beings are one animal species, products of nature. The natural and cultural evolution takes place within nature, on planet Earth. Naturalism entails, by definition, that no supernatural forces have had any effect on what has happened during our emergence or has any effect on what we do and think here and now. And all phenomena in nature are, in principle, within the reach of empirical inquiry. The question is what kind of inquiry? Is it natural scientific inquiry or some wider view of empirical investigation?

The basic idea of naturalism gets its expression in the principle of causal closure. Nature is causally closed and no other than causal forces are allowed in explaining what happens in nature. All events in nature take place through physical causal processes. One of the central problems is whether this entails that natural science is, at least in principle, enough for explaining all that happens in nature. Some authors seem to think so. John Dewey's naturalism is different in this respect. He sees science as a problem solving enterprise without any aprioristic commitments to specific methods. The character of the problem determines the choice of methods. Required methodological considerations are more general. Dewey's pragmatism contains radical criticism of the basic assumptions of the philosophical tradition.

P. Määttänen, *Mind in Action*, Studies in Applied Philosophy, Epistemology and Rational Ethics 18, DOI 10.1007/978-3-319-17623-9_1

1.1 Naturalism: Hard and Soft

Philosophical naturalism is generally associated with W.V.O. Quine's views. According to him epistemology is only a chapter in psychology understood as a natural science (Quine 1969, p. 82). Mind is reduced to the brain and mental concepts are "explained away" (Quine 1995, p. 86). This version of naturalism can be called hard naturalism because of its appeal to "hard" natural science. This move certainly makes all epistemological problems empirical problems—which is what one would expect of naturalism. But what justifies the claim that epistemology is a part of *natural* science? This seems to be a presumption that is just taken for granted. Dewey's view of inquiry as problem solving is not a commitment to the methods of natural science. It allows of methodological pluralism, which is why it can be called soft naturalism.

Commitment to naturalism entails that the aprioristic epistemology of classical philosophy is rejected. The problems of traditional epistemology either are solved by empirical methods or dissolved by showing that they are based on background assumptions rejected by naturalism. As pointed out, it does not follow that the empirical means should be those that are used in "hard" natural science. Dewey (LW 4) (1984) developed his operational conception of knowledge on the ground of analysing the development of physics, analysing the transition from Isaac Newton's theory to the theory of relativity and nuclear physics. However, the conclusion is *not* that physics with its own specific methods provides the ultimate explanation of all issues. There are similarities in the methods of inquiry but they are at a more general level (see his *Logic, Theory of Inquiry,* LW 12, 1938/1986). Inquiry is problem solving, and any method can be used if there is reason to believe that the use of the method gives information that is useful in solving the problem at hand. Dewey's conception of science as problem solving excludes the idea that one should cling to any specific methods, especially to those of natural sciences. Social sciences have different problems that may require different methods. It is not necessary to stick to the methods of natural science in order to keep problems and methods empirical.

In its effort to bring psychology under the scrutiny of natural sciences hard naturalism often ends up with reductionism. Folk-psychological predicates refer to nothing and are comparable to former scientific concepts like caloric fluid or phlogiston. This line of argument is problematic (Määttänen 2006). It is impossible to understand that a person is, for example, eating because she is hungry on the ground of neurophysiological terms alone. Actually there is not much to understand if this kind of reductionism is carried out. The subject matter of any psychology is human behaviour, and behaviour is defined in terms of the so-called folk-psychology. If these terms are removed, then there is nothing left to explain and understand. As Bennett and Hacker explain, mental predicates are attributed to people on the ground of their observable behaviour (Bennett and Hacker 2003). Mental predicates, observable behaviour and the understanding of the behaviour go hand in hand.

Apparently the need to get rid of folk-psychological (intentional and teleological) terms is partly motivated by the idea that these terms do not belong to the vocabulary of natural science. Dropping them away undeniably neatly dissolves the traditionally problematic character of these terms. However, the problems connected to these terms can also be solved if one removes hard naturalism instead (see Chap. 5). Dewey's soft naturalism gives the basis for doing it. Dewey's classical criticism of the reflex arc concept points to the fact that brain processes are not necessarily exactly those causal processes that realize cognition. The notion of the sensorimotor circuit, the loop of perception and action, gives the basis of the solution. Action and perception are realized through physical causal processes. There are no good reasons to a priori exclude causal processes external to the brain from cognitive processes.

The soft naturalism adopted in this book does not make any commitments to any specific methods of natural science. It is based on the idea that nature is causally closed. There are no absolutely a priori conceptual truths or methodological principles. All problems are at the final instance empirical problems. This statement seems to be moderate, but actually it entails a serious challenge to classical philosophy. This philosophical tradition is based on dichotomies and background assumptions that must be rejected in consistent naturalism.

The idea that all problems are ultimately empirical problems does not imply that the structure of experience adopted by classical empiricism is the appropriate epistemological stand to apply. However, some naturalists appealing to natural science seem to think so. Consider what Jonathan Knowles writes in a book dedicated to naturalism: "There is never a cognitive meeting between 'things-in-themselves'—that is, the things described by physics—and a similarly 'transcendental' organism" (Knowles 2014, pp. 213–214). This fits clearly with the classical view. Hidden causes of perceptions (things-in-themselves) are replaced by theoretical objects of science. Quine has adopted the same view. He states quite frankly (but erroneously) that Peirce is "an old empiricist" (Quine 1969, p. 78) and is content with Humean empiricism, as is clear already from his emphasis on the notion of stimulus meaning (Quine 1960). And "[o]n the doctrinal side, I do not see that we are farther along today than where Hume left us. The Humean predicament is the human predicament" (Quine 1969, p. 72). Quine fails to see the difference between classical empiricism and pragmatism, where action is included in the concept of experience.

But human beings are not Humean beings. There is a growing awareness of the need to change the views about the structure of experience as well as about the emphasis of hard naturalism. Tibor Solymosi refers to John Dewey and states quite correctly that we have to reconstruct our conceptions about experience, reject the classical spectator account of knowledge, and put emphasis on interaction (Solymosi 2013). Joel Krueger appeals to William James and stresses that conscious experience is action, not simply something that happens to us (Krueger 2006). And interaction with cultural environment is not necessarily best analysed with natural scientific methods. Hard naturalism is the dominant trend, but alternatives are beginning to emerge. John Ryder, for example, is quite explicit in taking distance to this version of naturalism (Ryder 2013, p. 20). Also Lawrence Cahoone takes the stand that "it is possible to formulate naturalism that, while incorporating the physical, is not physicalist, accepts reductive and non-reductive scientific explanations as complimentary, and coheres with the work of multiple sciences, hence is scientific yet pluralistic" (Cahoone 2013, p. 3).

1.2 A Priori Conceptual Analysis

Classical philosophy assumes that reason is independent of experience and capable of a priori conceptual analysis. This analysis gives timeless conceptual truths that are absolutely independent of how the world is and what is our experience of it. This idea has its origin in antiquity. Reason is a power to investigate, for example, what are Platonic unchanging ideas. The model was mathematics. Geometrical truths, or so it seems, are not recognized as such by experience. The sharp separation of reason from experience continued to later times (see, e.g., Hume 1978, p. 157). Aprioristic conceptual analysis takes it for granted that concepts (meanings) are formed and exist as vehicles of cognition independently of the world and how it is experienced. This is fine if, for example, it is assumed that consciousness is an immaterial container of immaterial units of thought and functions independently of the so-called external world.

From the naturalistic viewpoint this conception has a serious problem. If human beings are one species of animals, from where did we get this power of reasoning that transcends itself over and above nature? Naturalism does not allow of supernatural sources of cognition. We have got nothing "out of the a priori blue", to use John Dewey's phrase. All powers of cognition have developed within nature. This entails there is no absolutely a priori power of conceptual analysis, and this, in its turn, implies that the foundations of thought cannot be investigated only by thinking about thinking. Multidisciplinary empirical research is also required. All concepts are connected to various forms of experience. Of course, the connection is not direct; there is no reason to deny the power of abstract thought (or the ability to write pure fiction).

The classical view about a priori concepts would entail that linguistic meanings and concepts are independent of experience. This is not acceptable for a naturalist. The pragmatist notion of meaning is different. According to Charles Peirce meanings are habits of action. Action is something that embodied beings perform in nature. This pragmatist notion of meaning is completely consistent with the naturalistic principle that nature is causally closed. There is no need to postulate meanings, concepts or mental contents that would exist outside and independently of this causal closure. The use of linguistic expressions is one form of human practices interwoven with other practical activities, that is, experience. Nature and culture are not separate and independent of each other. There are no human practices that could transcend us over and above nature. Not a word is emitted without some bodily behaviour.

The pragmatist notion of meaning brings the reason as the use of symbols within the sphere of experience. The use of symbols takes place in the context of other practices within nature. However, a misleading dichotomy of culture and nature may lead to a conception according to which culture as symbolic activity indeed is over and above of nature and thus transcendental in a certain sense. Combined with a view that human cognition is basically determined by culture we get the false conclusion that cultural practices are independent of our experience as embodied beings. This is a neo-Kantian view. In Immanuel Kant's idealism pure forms of sensibility, pure concepts and categories of understanding exist absolutely a priori. They give the structure and the order of things in nature, which is constituted as an object of our experience by virtue of these features of human sensibility and understanding. Richard Rorty apparently thinks that language does the job. "But nature undescribed in any human language is simply Kant's unknowable thing-in-itself—an utterly useless notion, the plaything of philosophical skeptics, a toy rather than a tool" (Rorty 1997, p. 17). Then he sums up: "the more one thinks about language, the less need there is to think about nature" (ibid.). The question that he does not pose (let alone answer) is how we got the ability to use language in the first place. In Rorty's conception language does the job that in classical philosophy is given to the conceptual realm. But from where does the order of the conceptual realm come from? There seem to be two alternatives. The nature just is ordered in some way and our knowledge must accommodate to that order, or the concepts just are ordered in a certain way and the nature is carved up accordingly by some unknown mechanism that nobody has succeeded to describe in any detail. Those who think that the nature is carved up according our concepts should give some explanation how this takes place, and what concepts are independently of our already being products of nature. This seems an unsolvable problem because words don't gain meaning only by virtue of their mutual relations. Meaning is use, and the use of language takes place in the context of other practices with which we experience the world.

The conception that we are somehow inside language and cannot step outside of it is erroneous in that it neglects the role of the body in experience. Richard Rorty rejects explicitly the significance of embodiment by underrating those who slam a table and claim that they are expressing something significant with that. According to Rorty to slam a table is not a way to get outside language. Those who claim so are essentialists who, erroneously, think that they can have access to an object's intrinsic features. An antiessentialist can pick up an object only as articulated by a certain set of sentences (Rorty 1999, p. 56).

It is, however, a mistake to call a proponent of embodied experience an essentialist. By tapping some table one gets knowledge not about the table's intrinsic features but about certain objective and real conditions of action, that is, one comes to know that muscular effort meets resistance. And that's all there is to it. One way to express the idea is to say that by tapping a table one gets knowledge about *interactional* features instead of intrinsic features. It is hard to take this aspect of experience into account if one ignores our embodied existence and thinks that language is the only vehicle of thought and the only tool that gives structure and order to the experienced world. Naturalism entails that we are living organisms first, and the capacity to use symbols comes later (for one account of such development see Donald 2002).

Soft naturalism admits that symbolic practices have an irreducible and autonomous status but rejects the idea of a sharp dichotomy between the use of language and other practices. Symbolic practices are in many ways tied to other practices, which form the ultimate source of all meanings.

However, the notion of the a priori can be used in a relative sense. Konrad Lorenz wrote about biological a priori referring to the fact that, for example, in order to see colours we need eyes, biological sense organs. From the viewpoint of the experience of an individual organism this is before experience, that is, a priori. The organic body of living creatures is a prerequisite of experience. From the viewpoint of evolution our ability to see colours is a posteriori. Long evolutionary experience has produced organs necessary for seeing colours. Similarly the social and cultural system exists already and every new member of the human culture must adopt and accommodate to the existing way of using symbols, tools and other artefacts. Linguistic concepts are a priori from the viewpoint of each individual, but only in a relative way. In the final instance the human culture is a product of natural and cultural evolution. It is also based on evolutionary experience and is thus ultimately a posteriori. Also the methods of natural science have evolved during the progress of achieving knowledge, but once they are at hand, they are operationally a priori, as Dewey puts it (Dewey LW 12 1938/1986, p. 21). This relative use of this notion referring to evolutionary heritage is not what the representatives of classical philosophy had in mind. The inability to distinguish between these different senses of the notion of a priori leads to confusions. It is misleading to claim that only relatively aprioristic things have the power to give structure and order to nature. The body in the structured environment of everyday solid objects is not only the first instrument of experiencing the world. It is the necessary prerequisite of our existence and all cognition.

1.3 A Priori Laws of Thought

Classical aprioristic epistemology maintains that conceptual analysis gives eternal necessary truths that are absolutely certain and independent of how the empirical world is and how we experience it. How can this necessity be explained? From the viewpoint of naturalism, there can be no explanation of this by referring to abstract entities that are what they are, and philosophers consult them with the reason (separated from experience). This is a circle. Why are they what they are supposed to be? Who guards the guards? How do we know that these entities exist? They are simply thought into existence in order to explain the intuitively clear distinction between necessary laws of thought and empirically given laws of nature. However, naturalism cannot allow of any genuinely transcendental principles coming from over and above nature.

For a naturalist, the only way to investigate cognition is to start from the fact that human cognitive capacities (logic and mathematics included) are formed during the natural and cultural evolution down here on planet Earth. Human cognition is a system for controlling behaviour in natural and cultural environment. The principles governing cognition, the laws of thought, simply must have something to do with how the worlds is and what is our experience of this world, that is, natural and cultural environment.

Aristotle developed his theory of syllogisms on the ground of what he believed to be the world's general categorical structure. Also the classical laws of thought arise from this structure. The three classical laws of thought are the law of the excluded middle, the law of contradiction and the principle of identity. The law of the excluded middle says that a sentence is either true or false and nothing else. The law of contradiction says that a sentence cannot be both true and false. The principle of identity says that a thing is identical with itself. These laws have their ground on Aristotle's syllogisms, which are not independent of how the world is. Aristotle assumed that the world consists of individual objects and their classes. The class of all existing individual objects can be unequivocally divided to subclasses (and their subclasses until the complete and exhaustive categorization of all individual objects is reached) on the ground of essential properties that these objects either have or do not have.

In this view, there is a hierarchically organized structure of classes and subclasses that enabled Aristotle to investigate the validity of a syllogism (e.g., all animals are mortal, horses are animals, ergo horses are mortal) on the ground of its form. The form of a syllogism is that what is left when its content is dropped out. If we drop the talk about animals, mortality and horses we have a class, its subclass and an essential property of the class. All essential properties of a class are also properties of all of its subclasses. This is how we can investigate the validity of a syllogism on the ground of its form only. All syllogisms, which have the same form as the example above, are valid.

Now what about the laws of thought? The law of the excluded middle reflects the (assumed) fact that every individual object belongs to exactly one class of objects. "This is a horse" is either true or false, that is, the object cannot be partly horse and partly rose, for example. Similarly essential properties cannot be partial. An animal cannot be partly mortal. The law of contradiction reflects in the same way the structure of the classified world. If it is a horse, then it is a horse and cannot be a zebra, for example. More accurately, it cannot belong to any other class of objects. The situation is the same with the principle of identity. Horse is a horse; zebra is a zebra and so on. In short, the laws of thought are derivative of the (assumed) structure of the world.

Later on (with the favourable assistance of René Descartes) these logical principles were deprived of their connection with the world and raised to transcendental principles over and above nature. This is not to say that they are bad or wrong laws of thought. It entails only that there are some conditions of their successful application. Some parts of the world are neatly organized in exact compartments. Consider, for example, the periodic table of elements. One of the reasons why exact methods have proved to be useful in natural sciences seems to be the fact that nature is organized in an appropriate way so that certain principles of cognition can be successfully applied.

Wittgenstein (1956, pp. 31–34) considered the conditions of application of the rules for counting with the following example (it is modified a little). Suppose that there are 12 lines of 12 soldiers next to each other. The rules of counting tell that there are $12 \times 12 = 144$ soldiers. The soldiers are ordered to take a count. The first one says one, the next one says two, and so on. Suppose that the last one says 145. The first thing that comes into mind is that something went wrong. But suppose that this happens repeatedly. Does this imply anything for the rules of counting? No. The rules are what they are, but they could not be applied if the units one counts would really behave in this way. The obvious precondition for applying the rules of counting is that there are well defined and stable units to count. In nature there are such units, and the role of these units in animal life has probably something to do with why and how these rules have emerged.

Dewey (LW 12) (1938/1986) tried to establish the foundation of logic in a similar fashion as Aristotle, but there is one major difference. Aristotle based syllogism on the essential properties of individual objects. Classes of objects and their essences are fixed and immutable and therefore proper objects of knowledge. Logical necessity is based on generality, on the feature of nature that it just is categorized into classes on the ground of the general (essential) features of objects. As Dewey pointed out, examples of syllogisms containing individuals (e.g. Socrates is mortal) are slightly misleading because they appeal to particulars instead of essences (Dewey LW 12 1938/1986, p. 100). Logical necessity is based on general features.

A pragmatist cannot accept the view that the perceived world as consisting of classes of objects has general features affording the basis of logical necessity. In pragmatism the notion of experience is different. For Dewey the world consists of processes. Dewey's idea is that there are general features of inquiry as ongoing activity. Logical necessity is based on the generality of certain features of inquiry. Dewey's enterprise remained at a programmatic level. The important point to note here, however, is the fact that a naturalist cannot allow of any transcendent supernatural sources of logic and mathematics. Cognition as a whole is a capacity that evolved to human beings during natural and cultural evolution, and all "laws of thought" have their origin in this evolution. There are no other sources.

George Lakoff and Rafael Núñes have presented a theory according to which the ultimate source of mathematical cognition is found in the features of the embodied mind (Lakoff and Núñes 2000). Unlike Dewey, they don't pay much attention to external operations. For Dewey, symbolic operations are ways to act without acting, ways to anticipate the consequences of overtly performed operations (Dewey LW 4 1984, p. 120). Maybe the answer to the question of what is the ultimate source of the features of embodiment can be found by analysing the character and evolution of organism environment interaction with and without external instruments.

1.4 Empirical World—Apparent or Real?

The classical ideal of true knowledge is tightly connected with the dichotomy of the changing empirical world, which is only apparent, and the real being as the object of genuine knowledge, as something about which we can have unchanging timeless truths. The ultimate motivation for this distinction is precisely the need for some unchanging object of knowledge and timeless truths about it. There are two sources of this dichotomy that are worth mentioning here, namely ancient philosophy and the Cartesian separation of mind and matter.

Pre-Socratic philosophers had difficulties in conceptualising movement and change. For example, they knew that Achilles certainly runs faster than a turtle but did not understand enough the mathematics of infinite series. True knowledge, on the other hand, was assumed to be timeless, permanent and stable. And the object of true knowledge was assumed to be fixed and immutable. (See Dewey LW 4 1984). As they did not find anything of the kind from nature, they came to the conclusion that the perceived world, nature, is somehow unreal and only apparent. The real objects of knowledge, they concluded, are something else. Platonic ideas (or forms) are the best example of fixed and immutable objects of knowledge that cannot, by definition, be perceived. The only way to get knowledge about them is rational discussion. They can be accessed with thought. The idea of knowledge as unchangeable timeless truths can be saved if the ideas are taken as the real world that is independent of our empirical access to nature (sense perception). The mode of existence of these ideas is different than that of the empirical world. The motivation for this is the fact that such move helped to save the idea of genuine knowledge, immutable and eternal truths.

The dichotomy of apparent and real gives the alleged basis for claiming that the experienced world in which we live is somehow unreal and only illusory. The only possible grounds and motivation that the experienced world is somehow unreal is that there is some other realm of being such that is somehow "more" real. The problem of this dichotomy is that this realm of ideas as the object of true knowledge is, by definition, only thought into real existence. It cannot, by definition, be empirically accessed. The only alternative left is to say that we have epistemic access to the real being by virtue of pure thought. But we cannot simply think things into real existence and then claim that we can gain knowledge about these things with the same method, namely thinking about them. It is plainly circular.

Unfortunately the dichotomy of apparent and real is still going strong as the distinction between perceptions and their hidden causes. René Descartes invented the modern concept of consciousness (Bennett and Hacker 2003, pp. 12–23; Rorty 1980). Thereby he established the dichotomy of external and internal and mixed it with the dichotomy of real and apparent. Consciousness was separated from matter as something internal that gets to know the external material world via the sense organs. Perceptions are effects caused by external objects, and the epistemological problem is how to get knowledge of these unobservable causes of perceptions on the ground of perceptions. Interestingly, also Descartes argued circularly but inversely. He did not think things into real being but thought things out of real existence, namely his own body and the rest of the nature. He could doubt the existence of his body but could not doubt the doubt. From this he concluded that the doubt, *cogito*, could exist independently of the body. But this is wrong. The doubt can still be doubt exercised by a living body, as it in fact is. Mere doubt doesn't really annihilate the body. And if one admits that mind is embodied, then it does not make sense to doubt the existence of the so-called external world because the body is made of exactly the same stuff as trees, tables and galaxies. One would doubt one's own existence, which is silly. The body is a part of nature and that's all there is to it.

The other aspect of the dichotomy of internal and external is epistemological. If this question is addressed within the traditional framework, then it seems reasonable to ask questions like whether we are really brains in a vat. However, this question makes sense only on the assumption that knowledge can reside literally in the brain. Nobody has ever given any sufficient argument for this. It is simply assumed as a relic from the 1700th century. Pragmatist analysis of experience and knowledge and of the object of knowledge does not need any explication of the location of knowledge (or consciousness) literally in the brain.

The attempt to define the real objects of knowledge as something independent of our empirical access resembles the earlier conclusion to think fixed ideal objects into real being as genuine objects of knowledge. Both types of entities are by definition something that cannot be perceived. They are not objects of experience. The difference is that we are supposed to have privileged access to ideal objects as mental entities, but the hidden (material) causes of perceptions are difficult to access. David Hume asked the question whether perceptions are produced by external objects resembling them. The answer is, according to Hume, that "here experience is, and must be entirely silent" (Hume 1975, p. 153).

The problems and background assumptions that motivated the dichotomy of apparent and real are no good. Contemporary natural science has successfully conceptualised movement and change. Infinite series can be handled in mathematics. The original motivation for the dichotomy of real and apparent is gone. There are no good reasons to stick to the philosophical ideal of knowledge as eternal conceptual truths that are independent of how the world (nature) is and how we experience it. And there is not much evidence of disembodied consciousness wondering whether something external exists or not, and what could it be like. Consistent naturalism rejects the whole dichotomy. There is only one world and we live in it, in the web of various interactions. We interact with the world with our bodily organs. This is the world we perceive. Then we have different external instruments with which we measure and probe the world. With these technical means we interact with objects and properties that cannot be perceived with bare sense organs. These are really just two viewpoints to the world. The viewpoints are physical, not conceptual. This difference of physical viewpoints is not enough to support the classical idea that there "different levels of reality" of which one is "more real" than the other.

1.5 Ontological Symmetry

Philosophical naturalism denies all immaterial entities that are supposed to have some effect on the material world. Nature is causally closed. Ontological symmetry is one consequence of this. It follows from naturalism that the mind is necessarily embodied. The living body exists as a part of nature, which makes life possible. However, there remains an obvious epistemological asymmetry. Some parts of nature have experience and knowledge about other parts of nature. The problem is the character of the embodied mind as thinking and knowing subject. For a naturalist the problem of the existence of the so-called external world is dissolved as a direct consequence of the principle of causal closure. In causally closed nature there is no room for immaterial entities. The questioning of the existence of the material external world presupposes an immaterial sceptic, but there is no such thing. Serious scepticism about the existence of the material world is not a meaningful stand in naturalism.

The ones who know are biological organisms in nature. They are made of exactly the same stuff as the world that is external to the body. The structure of the brain just is, luckily, more complex than the structure of trees, for example. But ultimately the material consists of same kind of elementary particles. If one knows that she exists as a live creature, she knows with equal certainty that the natural environment exists. Note that to know *that* something exists is not the same as to know *what* it is. Further, ontological symmetry is compatible with epistemological asymmetry. Organisms with brains gain knowledge about trees and not the other way round. The problem is what we can know about ourselves, about this environment and about our relationship to this environment.

Naturalism entails that mind is embodied but it does not necessarily imply that mind must be reduced to the brain. The brain is the organ of thought but it is not the brain that thinks. A human being thinks with the brain. Just like the legs are the organs of running but it is not the legs that run. A human being runs with the legs. To say that the brain can think just by itself would be the same as to say that a pair of legs cut off from a body can run. The brain functions properly as an organ of thought only in the context of the interaction of living organisms and their environment. The problem whether we are just brains in a vat is meaningful only on the assumption that knowledge can literally reside in the brain, but nobody has satisfactorily explained what it would be for the brain to think and contain knowledge just by itself, independently of the rest of the world. The dichotomy between external and internal needs to be challenged once and for all.

Mental states must be realized through causal physical processes, but there is no compelling reason to maintain that these mental states are internal in the sense that they are literally located within a living body. One reason for this is that they are not individual units that can have a location in the same sense as material objects like chairs and tables. But as we shall see in Chap. 5, thoughts are not individual units but relations. And relations don't have location in the same sense as individual objects. Suppose a table is bigger than a chair. The table and the chair are located on the floor. But where exactly is this relation "is bigger" located? This is not a good question. Similarly, thoughts are relations between the brain, body, and the physical and social environment. Cognition is realized by a complex system of relations between human beings and the rest of the world. These relations don't have locations in the same sense as individual objects have. Especially they are not located in the brain. The brain is an element in this system of relations, but it would be a logical category error to reduce a relation to one of its elements. The relation simply vanishes by this kind of reduction.

1.6 Naturalism and Normativity

One of the misleading dichotomies of classical philosophy is the one of nature and culture. This dichotomy suggests that human reason and language is the distinguishing feature of human that raises us over and above nature. Especially morality, values, and norms arise in human society. Nature is the realm of pure causality where norms simply do not exist. On this ground it is claimed that norms cannot be naturalized. The same holds for goals and purposes. "There are no purposes in nature; physics has ruled them out, and Darwin has explained them away" (Rosenberg 2014, p. 25). However, in John Dewey's soft naturalism things are different. He tried to give ethics a scientific justification and said that biology is the bridge between nature and culture (Dewey 1988, pp. 246–247). By building a bridge between nature and culture we can also overcome the unjustified gab between facts and values. And contrary to what Alexander Rosenberg claims, a concept of genuine intentionality is perfectly possible in naturalism (see Chap. 5). It is true that evolution has no goal, but this does not imply that we have to say farewell to the purpose-driven life, as Rosenberg believes (Rosenberg 2014, p. 27). Jerome Popp is of different opinion. "It should now be clear how evolutionary theory provides a scientifically created platform upon which one can build a theory of inquiry capable of answering our normative questions" (Popp 2007, p. 93). This difference in opinions is related to their versions of naturalism. Popp refers to John Dewey, but Rosenberg seems to be trapped in hard naturalism. Evolution does not have goals, but this does not imply that individual organisms and their groups cannot have goals.

The basic reason for a misleading separation between nature and culture as well as between facts and values is the erroneous conception experience as mere sense experience. The sense organs are, in this view, sort of channels that connect the internal mind to the external world. However, in pragmatism the concept of experience is wider (Chap. 2). Our interaction with the world consists of perception and action, and both of them have an epistemic role. The world is experienced as possibilities of action (affordances), and the object of experience and knowledge is the relationship between two experienced situations: the present situation here and now and the future situation that is an outcome of some way of acting.

There are always a large number of possibilities in every ordinary situation. And it is practically impossible to carry out all possibilities of action at one time. Therefore one is literally forced to make some choices between the alternatives. This is, in effect, valuation. In every step and move we make, we are valuating (consciously or subconsciously) anticipated outcomes of possible ways of behaving. The basis of these choices is the relationship between expected experiences and current needs and desires. This entails that values are present in the experienced world. An acting agent is necessarily a valuing agent.

The fact of a pragmatist is radically different from the fact of an empiricist. Humean empiricism takes facts to be that what one literally perceives here and now, and since values cannot be literally perceived they do not belong to the world of facts. If you don't literally perceive values (or causality), they don't belong to the world of facts. For a pragmatist facts are relationships between what is experienced now and what will be experienced in the future as an outcome of one way of acting. These future experiences are necessarily (often subconsciously) valued. Values are thus inseparable elements of the experienced world. Experience consists of perception and action. Therefore perceived facts and anticipated values (that are anticipated to be realized in the future as an outcome of chosen activity) are necessarily interwoven in experience. From this point of view, normativity is not a problem in naturalism. On the contrary, naturalism provides a natural solution to the problem of facts and values.

1.7 The Pragmatic Maxim of Charles Peirce

Philosophical naturalism entails that all human activities are performed within nature. This holds also for reasoning, the use of language and other symbolic resources. We have no real access to anything supernatural. This is one way to carry out consistently the naturalistic principle that the world is causally closed. Peirce's pragmatic maxim is as follows: "Consider what effects, that might conceivably have practical bearings, we conceive the object of our conception to have. Then, our conception of these effects is the whole of our conception of the object" (CP 5.402). This entails, among other things, that we can reach with our concepts and meanings only those features of the world that fall within the sphere of our possible experience. The appeal to practical bearings does not mean any restrictions to thought experiments, imagination and so on. The maxim "makes conception reach far beyond the practical. It allows any flight of imagination, provided this imagination ultimately alights upon a possible practical effect; and thus many hypotheses may seem at first glance to be excluded by the pragmatical maxim that are not really so excluded" (CP 5.196).

Meanings and concepts apply to the world as it is experienced. Early pragmatists were quite clear about this. The pragmatic maxim entails a rejection of such alleged objects of knowledge that are in principle, by definition, beyond the scope of our epistemic access to the world (see Chap. 2). Charles Peirce characterized Immanuel Kant as a pragmatist, although "a confused one" (CP 5.525). The confusion consists of the fact that after deciding, correctly, that the concepts of understanding can only be applied to experience Kant still employs the concept of the thing in itself, which is Kant's version of the classical real being that is in principle beyond the scope of experience. This holds also for experience mediated by instruments and experimental devices (see Sect. 7.2 dealing with instrumental phenomenology). If concepts can be applied only to possible experience, then so be it. Entities cannot be simply thought into real existence. We live in nature, and nature is the object of our experience and our knowledge. This fact forces us to revise the classical conceptions about experience and knowledge.

The pragmatic maxim can be used in philosophy of mind for rejecting empty disputes. One such issue is the thought experiment about zombies according to which some of us are not genuine human beings with internal mental states. They have no such states but are zombies that behave in a similar manner in all respects. If it really is the case that the alleged zombies behave like genuine human being in all respects, then we just cannot tell who is a zombie. This definition of a zombie make is impossible to the difference between a human being and a zombie to have any practical bearings in our experience. Therefore it is an empty problem.

Rorty (1980) has a similar strategy when he asks us to think about Antibodeans living on a remote planet. They are similar to us but there is one difference. They are not used to report their activities by saying things like: I am eating because I am hungry. They say instead: I am eating because my C-fibres are firing. The point is that there is no practical difference whether the cause of action is said to be internal mental state or internal neural activity. The whole difference is empty. Rorty's conclusion is that we can drop the talk about internal mental states altogether. There is no "mind stuff". There are no universals. In Chap. 5 we shall see, that in Peircean pragmatism it is possible to keep the mind stuff, real generals, but they are not internal entities.

References

Bennett, M., & Hacker, P. (2003). *Philosophical foundations of neuroscience*. Oxford: Blackwell.

Cahoone, L. (2013). *Orders of nature*. Albany: SUNY Press.

Dewey, J. (LW 4) (1984). The quest for certainty. In J. A. Boydston (ed.), *The later works 4*. Carbondale and Edwardsville: Southern Illinois University press.

Dewey, J. (LW 12) (1938/1986). Logic: The theory of inquiry. In J. A. Boydston (ed.), *The later works 12*. Carbondale and Edwardsville: Southern Illinois University press.

Dewey, J. (1988). Theory of valuation. In J. A. Boydston (ed.), *The later works 13* (pp. 191–250). Carbondale and Edwardsville: Southern Illinois University press.

Donald, M. (2002). *A mind so rare. The evolution of human consciousness*. New York: Norton.

Hume, D. (1975). Enquiries concerning human understanding and concerning principles of morals. In L. A. Selby-Bigge (ed.), Oxford: Clarendon Press.

Hume, D. (1978). A treatise of human nature. In L. A. Selby-Bigge (ed.), Oxford: Oxford University Press.

Knowles, J. (2014). Naturalism without metaphysics. In K. R. Westphal (ed.), *Realism, science, and pragmatism* (pp. 200–218). New York, London: Routledge.
Krueger, J. (2006). James on experience and the extended mind. In J. Shook & P. Ghiraldelli (eds.), *Contemporary pragmatism 3:1* (pp. 165–176). Amsterdam: Rodopi.
Lakoff, G., & Núñes, R. (2000). *Where mathematics comes from how the embodied mind brings mathematics into being*. New York: Basic Books.
Määttänen, P. (2006). Naturalism: Hard and soft. In H. J. Koskinen, S. Pihlström, R. Vilkko, & P. Lang (eds.), *Science—a Challenge to Philosophy? Proceedings of the xv Internordic Philosophical Symposium, Helsinki, May 13–15, 2004* (pp. 227–236). Frankfurt am Main.
Popp, J. (2007). *Evolution's first philosopher: John Dewey and the continuity of nature*. Ithaca: SUNY Press.
Quine, W. V. O. (1960). *Word and object*. Cambridge: The MIT Press.
Quine, W. V. O. (1969). *Ontological relativity and other essays*. New York: Columbia University Press.
Quine, W. V. O. (1995). *From stimulus to science*. Cambridge: Harvard University Press.
Rorty, R. (1980). *Philosophy and the mirror of Nature*. Oxford: Blackwell.
Rorty, R. (1997). *Truth, politics and 'post-modernism'*. Assen: Van Gorcum.
Rorty, R. (1999). *Philosophy and social hope*. London: Penguin Books.
Rosenberg, A. (2014). Dissenchanted naturalism. In B. Bana & D. M. Hans (eds.), *Contemporary philosophical naturalism and its implications* (pp. 17–36). New York: Routledge.
Solymosi, T. (2013). Neuropragmatism on the origins of conscious minding. In Swan (ed.), *Origins of mind* (pp. 273–287). Dordrecht: Springer.
Wittgenstein, L. (1956). Remarks on the foundations of mathematics. In G. H. von Wright, R. Rhees & G. E. M. Anscombe (eds.), *Bemerkungen über die Grundlagen der Mathematik*. Oxford: Blackwell.

Chapter 2
Experience and the Object of Knowledge

The Cartesian conception of mind is based on the dichotomy of external and internal, which, in its turn, is closely related to the classical conception of experience. Therefore, a consistent redefinition of mind in pragmatism requires a revision of the notion of experience formulated by early modern philosophers. In this view, experience is sense experience. Sense organs function like channels connecting internal consciousness to the so-called external world. Perceptions are caused by external things but these external causes itself cannot be perceived. Perceptions are effects of these hidden causes. The object of knowledge is the external world as it is independently of what we say or think about it and independently of our epistemic access to (perceptions of) the world. Sense perceptions are internally conditioned which is why they do not directly inform us about the real character of the world. This is because, among other things, effects don't resemble their causes. For example George Berkeley wondered how colours could resemble something than cannot even be seen.

The classical view of experience can be questioned. The same holds for the dichotomy of internal and external as well as that of apparent and real. The founders of pragmatism challenged classical philosophy by broadening the concept of experience. The role of action in experience must be taken into account. This, in its turn, leads to a radical redefinition not only of the concept of experience, but also of the notion of the object of knowledge. The hidden causes of perceptions are replaced by the (at present) hidden consequences of action as the ultimate goal of what has to be known. To know is to know what to do in order to achieve one's goals. Further, the difference between mere belief and true knowledge (absolutely a priori truths) is not tenable. In Dewey's operational conception of knowledge the earlier distinction between mere belief and true knowledge is replaced by the distinction between what is had and what is known.

P. Määttänen, *Mind in Action*, Studies in Applied Philosophy, Epistemology and Rational Ethics 18, DOI 10.1007/978-3-319-17623-9_2

2.1 Classical Conception of the Object of Knowledge

Richard Rorty has a point in his claim that the idea of a theory of knowledge grew around the problem of the "external world", that is, "the problem of knowing whether our inner representations were accurate" (Rorty 1980, pp. 139–140). The first and most radical problem concerns serious ontological scepticism: how do we know that the external world even exists. In its original Cartesian form the external world is material and the internal world is immaterial. In formulating his new conception of mind Descartes changed the meaning of the notion of idea. Platonic ideas are forms, but for Descartes the knowing subject is a thinking and unextended thing. Ideas in such a mind cannot be forms because form requires extension, spatiality. The epistemological problems become acute.

The Cartesian conception of mind is based on a container metaphor (Lakoff and Johnson 1999). Ideas are in the mind like cookies in a jar. People have a privileged access to their own consciousness by introspection. It became a specific epistemological problem to explain how the internal universal ideas can be knowledge of external particular objects. Ideas are universal and intentional internal units that refer to (are about of) external material objects. How are ideas and objects related to each other while they are so different things, and how can one universal idea be about all its objects that are also different in many ways? Locke based his answer on similarity, or conformity, to be more accurate. Idea and its object have similar form (a primary quality in contrast to secondary qualities like colours, for which no account of the connection can be given). The idea of circle is round and the epistemological relation between the idea and circles in the world is based on the same form, there is conformity between them. However, Locke could not solve the problem of a general triangle that should be "neither oblique nor rectangle, neither equilateral, equicrural, nor scalenon; but all and none of these at once" (Locke 1959, p. 274; see Määttänen 1993, pp. 21–30). This general triangle is required because of the demand of conformity between the idea and all these different triangles in the world.

This sharp separation of the mental (or intellectual) and the material (or sensibility) was a problem also for Immanuel Kant who developed his doctrine of schematism for explaining the connection (Määttänen 1993, pp. 21–29). Kant's Copernican revolution changed conceptions about perception but the general framework of experience remained the same. The hidden causes of perception, about which experience is silent as David Hume said, are changed to the thing it itself. The character of perception is, however, changed. The main idea of the Copernican revolution is that all right, we cannot get from effects to their causes, but there is no need to. Instead we can get from causes to their effects. This is related to the method of analysis and synthesis that has ancient roots. In the Middle Ages phenomena were investigated by searching their causes by analysis, and synthesis was the opposite procedure. Phenomena were produced by manipulating their causes or constructed (synthesized). This lead to the principle that one can get certain knowledge of things if one have access to their causes. For example Thomas Hobbes and Benedict Spinoza appealed to this principle (Hobbes 1962, pp. 3, 10–11; Spinoza 1955, p. 34). Kant applied the notion of synthesis to experience. Pure synthesis gives the manifold (das Mannigfaltige) a priori (Kant KdrV, A 77/b 103). Certain knowledge about nature can be achieved because nature as an object of experience is a product of the synthetic activity of pure understanding.

The outcome of all this is that to perceive is not to receive passively impressions from the world. Perception is an active and constructive process. This entails the character of the perceiving mind has an effect on what is perceived. Thus in order to find out what the experienced world is like one has to find out first what the mind is like and what is the mind's affect on perceptions. This is one important task of aprioristic epistemology. The need for such an epistemology is motivated by the view of perception as an outcome of the effect of the thing in itself and internal conditions.

The classical tradition tempts one to say that dependence on internal conditions can be termed mind-dependence. For early modern philosophers mind was immaterial consciousness. After Kant these internal conditions are often called concepts. Neo-Kantians tends to say that the experienced world is carved up by concepts (see for example Putnam 1981). The question that remains unanswered is what is the concept of concept applied in this context.

The Cartesian view of mind and its contents was mediated to contemporary psychology by Franz Brentano who made the unfortunate analogy between language and mind (Brentano 1924, pp. 124–125). Medieval philosophy of language had come to the conclusion that the words "horse" and "centaur" function in language in the same way quite independently of the fact that there are no centaurs. The words are intentional units that purport to refer to their intentional objects. The words are about something. Intentionality is *aboutness*. In a similar manner, mind consists of intentional units that are about something. Contemporary talk about mental or internal representations continues this tradition with respect to the dichotomy of internal and external. Dependence on internal conditions is dependence on internal mental states (internal representations, meanings or concepts). And intentionality became a criterion of mentality. It is a good criterion, but there is a better way to explain what it amounts to.

The classical view of experience and of the object of knowledge has many flaws. Once the role of bodily organs in perception is accepted the dependence on internal conditions becomes more complicated. Some internal conditions are material conditions, for example biological properties of the sense organs and the brain. Consider colours. In order to see the table as brown we need light, the table that reflects light and the eyes belonging to a living organism. Perceived colours are the joint outcome of these factors. Therefore it is justified to say that colours are properties of the concrete interaction of the live organisms and its environment although colours are experienced to be attributes of the perceived object. There is no reason to refer to hidden and thus unknown causes of perceptions. All factors involved are known well enough. We perceive the table as a material object in spite of the fact that we perceive it as coloured by virtue of having eyes. The table is the cause of its being perceived. The hidden causes can be removed from the discussion, and this removes also the need of any specific a priori epistemology to explain how we can gain knowledge about something that is in principle hidden from us. It may be argued that the table as a swarm of elementary particles is a hidden cause of perceptions, but this really does not change anything. Elementary particles are observed with external instruments, and observation with external instruments can be analysed in the same way.

In addition to material conditions there are, of course, conceptual conditions. Perceptions are interpreted with linguistic meanings, concepts and theories. Accordingly we can distinguish between two kinds of viewpoints, physical and conceptual. Concepts and theories change and conceptual viewpoint changes, we get new interpretations of issues. But physical viewpoint is different. Mind is inseparable from the body, and the body determines one's viewpoint to the world. Material conditions that determine the physical viewpoint are difficult to change. Biological evolution changes bodily organs, and external instruments that mediate and modify perceptions can be developed. Physical viewpoint is also objective in the sense that bodily organs are exactly as objective elements in the physical world as the physical objects perceived.

The possibility of radical ontological scepticism is one consequence of the classical doctrine of the object of knowledge combined with the notion of immaterial mind. How can one know that these hidden causes of perceptions even exist? The traditional sceptical problem is sometimes reformulated in contemporary terms in papers like Are We Brains in a Vat? (Putnam 1981). In this setting a naughty scientist has put the brain in a vat and manipulates its nerve endings. The brain is supposed to ponder whether it can know for sure that it is not just a brain in a vat. This question does not presume any immaterial mind, but it presumes that the brain is able to do the same things as the Cartesian soul. However, strictly speaking nobody has ever managed to show how knowledge (meanings, concepts, ideas, you name it) can literally be located in the brain, how the brain can just by itself take care of cognition. The so-called neural correlates of consciousness found by brain imaging are not enough to show this. This discussion takes for granted the classical distinction between external and internal, and this starting point seems to be a mere presumption inherited from the philosophical tradition rather than an established fact.

The brain-in-a-vat discussion is also a striking example of how the role of the rest of the body and the bodily action is simply ignored as if it had nothing to do with experience, cognition and knowledge. There are more and more evidence for the significance of the body and motor activity. Naturalism does not imply that mind must be reduced to the brain. Philosophical pragmatism with its emphasis on the significance of action changes the notion of experience and offers an alternative line of inquiry.

2.2 The Role of Action in Experience

According to Peirce the concept of experience is broader than that of perception. Experience includes much that is not perceived. "It is the compulsion, the absolute constraint upon us to think otherwise than we have been thinking that constitutes experience. Now constraint and compulsion cannot exist without resistance, and resistance is effort opposing change. Therefore there must be an element of effort in experience; and it is this which gives it its peculiar character" (CP 1.336). Effort and resistance are experienced in action. To experience is to be an active agent in the world. In a similar manner John Dewey criticized what he called a spectator theory of knowledge. As living creatures we perceive and act.

Now we face the question of the relation between perception and action. They cannot be simply separated as different elements in experience. They function together at the same time. Peirce characterized the difference as follows. In action "our modification of other things is more prominent than their reaction on us" while in perception "their effect on us is overwhelmingly greater than our effect on them" (CP 1.324). Precisely because of this difference action not only broadens the concept of experience but also changes its character.

Effort meets resistance that is compulsive. Hard facts resist our will. Peirce describes the character of hard facts with a sceptic walking down Wall Street "debating himself the existence of an external world; but if in his brown study he jostles up against somebody who angrily draws off and knocks him down, the sceptic is unlikely to carry his scepticism so far as to doubt whether anything beside the ego was concerned in that phenomenon" (CP 1.431). Hard facts make them to be recognized in experience, and in order to experience this one must act and meet the resistance of hard facts.

The resistance of the world forces one to accommodate one's activity to the hard facts, which function as objective conditions of action. The body is the first and necessary instrument for experiencing the world. Muscular effort meets the resistance of the physical world. Bodily behaviour accommodates to these objective conditions. To some extent the physical world can be changed by one's own effort, but this has obvious limits. We can change these conditions by using external instruments like tools, machines and other devices, but the same relation between effort and resistance remains. This is why action as an element of experience is more forceful than perception alone. "The authority of experience consists in the fact that its power cannot be resisted" (CP 7.437). This authority is experienced when we act in the world.

There is a difference between action and perception, but they function together. This entails that action plays a role also in perception. Classical empiricism maintained that perception is passive. Sense organs only receive impression from the world. This view has changed. Internal conditions have an effect on how the world is perceived. It is a commonplace that perceptions are interpreted with meanings, concepts, beliefs, theories and so on. The lesson to learn from pragmatism is that action is involved in the ways we perceive the world. This is observed also in contemporary cognitive science. According to Alva Noë what we perceive is determined by what we do or what we know how to do (Noë 2004, p. 1).

Peirce coined the term *percept* for expressing this idea. Percept is not the same as perception, which in the classical tradition is considered to be in the mind. There is a double awareness involved in percepts (CP 7.625). A table, say, is perceived to be brown and round. Perception is interpreted with meanings; it is internally conditioned. Objects as perceived, as cognised in the sign, are immediate objects (CP 8.183). This is one aspect of percepts. On the other hand, there is an element of hardness in percepts. The hardness of fact "lies in the insistency of the percept" (CP 7.659). Percepts resist our will. Percepts are perceived objective conditions of action.

In other words, the table is experienced as an object of perception, as immediate object. But at the same time it is experienced as an object of (potential) action where muscular effort meets resistance. The awareness of this on the ground of mere perceptions is based on earlier practical experience of dealing with rigid objects like tables. This earlier experience gives grounds to anticipate what will happen if some muscular effort takes place. As we shall see later, habits of action are formed when action is accommodated to objective conditions of action. These habits are meanings (Chap. 4). The hardness of percepts is based on the fact that they are interpreted with these non-linguistic (or tacit) meanings. It is the same table that is, at the same time, an object of perception and an object of motor action. The table is made of the same stuff as my body; both are equally objective and real parts of the material world. The immediate object, the table as perceived, is the very same object that resists muscular effort as a material object. Peirce goes as far as to say that, rightly understood, it is correct to say that we "directly perceive matter" (CP 1.419). Once the principle of ontological symmetry is accepted it is not problematic to say that we perceive matter.

So it makes sense to say that we perceive more than what classical empiricism would allow, but on the other hand Peirce claims that we experience something that cannot, properly speaking, be perceived. "We perceive objects brought before us; but that which we especially experience—the kind of thing to which the word 'experience' is more particularly applied—is an event" (CP 1.336).

Now what is an event? Event is a temporal process that involves some kind of change. A situation, say $\mathbf{S_1}$, is changed to another, say $\mathbf{S_2}$. There are events that take place in our environment without our participation, but and active agent is interested to know what to do, what events can be brought about. Action can be described as an operation or a set of operations (**O**) that connects these situations. The second situation is an outcome of some operations performed in the first one. Thus we get the scheme: $\mathbf{S_1} \rightarrow \mathrm{O} \rightarrow \mathbf{S_2}$. Both situations can be perceived but not at the same time (the same holds for events that occur by themselves). This is why events are not properly speaking perceived. An agent in the first situation perceives the situation but that what is of interest are the possibilities to bring about changes. The world may be *perceived* as phenomenal qualities attributed to perceived objects, but properly speaking it is *experienced* as possibilities to bring about changes, or better, as possibilities of action. In pragmatism to know is to know what to do. Spectator theory is replaced by operational conception of knowledge. Experience is, generally speaking, orientating to possible future experiences on the ground of past practical experience.

2.3 The Object of Knowledge in Pragmatism

Pragmatism entails a radical change in the notion of the object of knowledge. The goal of knowledge is not to reach the "real" but hidden and mind-independent world causing our perceptions. Hidden causes of perceptions are replaced by the hidden (that is, in the situation $\mathbf{S_1}$ hidden) consequences of action. As Dewey put it, the objects of knowledge are controlled processes of change where acting agents transform a situation into another (Dewey LW 1 1981, p. 128). Acting agents bring about controlled processes of change. The knowing subject and the performed operations belong to the object of knowledge. Thus we have two different senses of the notion of object. One notion is the world (or some part of it) as an object of perception and action and the other is the object of knowledge that is a relation between two situations. This relation is mediated by controlled operations performed by the knowing subject: $\mathbf{S_1} \rightarrow \mathrm{O} \rightarrow \mathbf{S_2}$.

What will be known is the joint outcome of some processes in the world observed in the first situation and some operations that a started in this situation. The outcome is the second situation. What will *not* be known is how the world is as alleged hidden causes of perceptions before the operations. Experience consists of perception and action, and the limits of experience are relative to the methods and instruments of investigating the world (Chap. 7). The difference between the classical view and the pragmatic view as to the perceived situation $\mathbf{S_1}$ can be described as follows.

The classical view of perception as the (only) epistemic access to the world requires that there are internal items (in some forms of naturalism neural events in the brain) that function as internal representations and that there is an epistemic relation between an internal representation and the perceived object. The problem is to explicate by virtue of what is a neural event in the brain a representation. By virtue of what is it (maybe in combination with other neural events) knowledge? What makes a neural process a mental process, an intentional entity? These problems are addressed later. They are far from solved.

Pragmatism emphasizes that knowledge is prospective, not retrospective. An active agent seeks ways to act further and wants to know what to do in order to reach the situation $\mathbf{S_2}$ as an outcome of this action. The second situation is hidden in the first but it will be revealed if knowledge is adequate. Thus knowledge is always relative to the (actual or potential) activity of the knowing subject. This is the way in which things are even in situations where there are no conscious objectives of action. Beliefs (habits of action) need not be conscious, as Peirce already noted (CP 2.148, 2.711 and 5.417). Mere observation may seem to be static and instantaneous, but this is illusory. Simple perceptions are complex and fast operations. Action is in a way involved also in perception (Noë 2004). To know that snow is white is to know how to look at snow and recognize its whiteness as a result of a complex and active process. The problem that remains to be solved is the question of what is the role of internal neural events in experience.

The structure of experience and the object of knowledge can be described with the following figure.

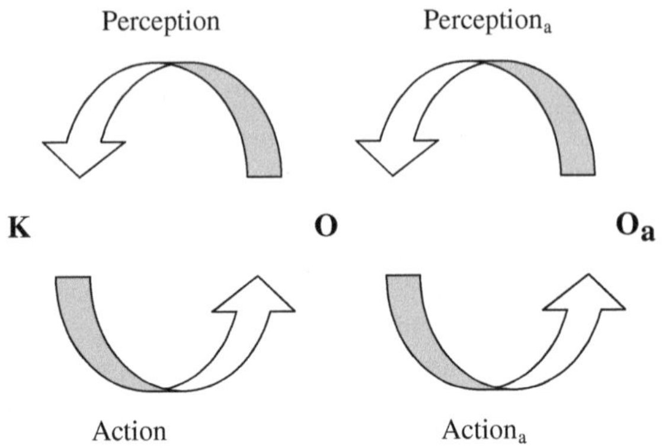

The knowing subject **K** faces the world as an object of perception and action. The world is experienced as possibilities of action. Each situation offers a large number of possibilities of action. These possibilities are anticipated with earlier acquired habits of action. Anticipatory mechanisms (Sect. 3.4) bring to mind what are the expected consequences of anticipated action (Action$_a$), what would be perceived (Perception$_a$), namely the anticipated object of experience (**O**$_a$).

The world is an object of perception and action, but the object of knowledge in pragmatism is a different notion. It is a relation between two situations, subject encountering **O** (situation S_1) and subject encountering **O**$_a$ (situation S_2). In Dewey's words: "The objects of science, like the direct objects of the arts, are an order of relations which serve as tools to effect immediate havings and beings" (Dewey LW 1 1981, p. 110). The relation between situations is mediated by action, and this brings the activity into the object of knowledge. This view differs radically from the classical view. One important difference is that there is no need for epistemology the task of which would be telling a priori (in the absolute sense of the word) how the character of mind (as an internal knowing subject) affects the perceptions about the so-called external world. All prerequisites of having and acquiring knowledge are formed on the basis of past practical experience. Past experience consists of evolutionary experience, cultural heritage and individual experience (those elements of evolutionary and cultural experience that an individual succeeds to acquire during growth, socialization, education and so on). Another difference is that the acting agent (knowing subject) belongs to the object of knowledge. This follows from the fact that action is needed in order to change the situation. The knowing subject is involved or embedded in the situation she is transforming by acting in it.

2.4 The World as Had and the World as Known

According to Dewey "we do not have to go to knowledge to obtain an exclusive hold on reality. The world as we experience it is a real world" (Dewey LW 4 1984, p. 235). Peirce emphasized the role of the resistance of the hard facts as objective conditions of action. This world is real and it is an object of perception and action, and the object of knowledge is a relation $\mathbf{S_1} \rightarrow \mathrm{O} \rightarrow \mathbf{S_2}$.

Dewey makes use of the distinction between had and known. When the problematic situation $\mathbf{S_1}$ is encountered, it is simply had as an object of perception and action. It is a situation that one begins to inquire. During the inquiry it begins to change into one element of the object of knowledge. $\mathbf{S_1}$ is experienced as possibilities of action that have anticipated consequences. The problem determines what kind of consequences are the desired ones. This, in its turn, determines what kinds of operations are required. These operations will be found on the ground of the relevant features of $\mathbf{S_1}$, general knowledge about the world as it is experienced in $\mathbf{S_1}$ and previous experience of similar situations. Now we have: $\mathbf{S_1} \rightarrow \mathrm{O}$. If the selected operations turn out to be successful, then we have the object of knowledge $\mathbf{S_1} \rightarrow \mathrm{O} \rightarrow \mathbf{S_2}$. We have been justified in claiming in $\mathbf{S_1}$ that we know how to solve the problem, how to proceed to $\mathbf{S_2}$.

Dewey illustrates his conception with an example of a patient coming to see a physician (LW 4 1984, pp. 139–140). The patient sets the problem of inquiry. Some features of the patient are searched out as symptoms. These symptoms are the relevant data for making a diagnosis with the help of theoretical knowledge about medicine and previous practical experience of treating patients with similar symptoms. Some operations are performed to make the person healthy again. If this really happens, then the physician has had knowledge adequate enough about the problems in the health of the patient. Knowledge is prospective in the sense that the adequacy of knowledge depends on the course of events during the treatment. The knowledge is, of course, based on earlier experience and theories involved, but if the treatment is not successful, then the knowledge turns out to be inadequate and, thus, after all not true knowledge about this problematic situation and its transformation into the desired situation. An acting agent wants to know what to do and orientates to the future on the ground of past experience, but the proper justification of knowledge takes place in the future. Knowledge is adequate enough if the action performed turns out to be successful.

The transition from what is had to what is known makes Dewey's use of the word "object" quite flexible. Some objects of experience are simply had, and then there are objects known. Hasty interpretation might consider these objects as distinct objects. However, this interpretation is not quite correct. The patient coming to the physician is first simply had. She has various features. Then the problem, the fact that the patient has some symptoms, starts the inquiry (examination, laboratory test and so on). When a treatment has been decided, the patient has been transformed into an object known. The person has become an object of medical knowledge. But it is the same person observed and interpreted in the framework of medical science. The person as had and the person as known (as an element in the object of knowledge) are not two distinct objects. The same object is interpreted with different sets of meanings. So far as these meanings are considered to constitute the object we have two objects, but only one person.

In pragmatism knowledge is not necessarily unequivocal. There are often many ways to reach the goal, and the criteria of evaluating these different ways may depend on the context. Further, the goal may be reached accidentally, or inadequate action may lead astray. If the first situation remains problematic without any clue of how to act adequately, then the situation remains as an observed situation that is simply had and does not belong to the object of knowledge. This distinction between the world as had and the world as known is a distinction between true knowledge and mere opinion. It is a parallel to the classical distinction between knowledge as justified and eternally true knowledge about the real being as distinguished from mere opinions about the moving and changing empirical world. In naturalism there are no good reasons for making such distinction. There is only one world and we live in it. What is left is a distinction between had and known.

The view, according to which the object of knowledge consists of the hidden causes of perception, is still with us. The apparent plausibility of this view is based not only on the inertia of the classical tradition in philosophy, but also on contemporary philosophy of science. The table as a perceived object and the table as a swarm of elementary particles are sometimes considered as two different things. Elementary particles cannot be seen, so the idea is that these theoretical objects of science are the hidden causes of sense perception. In a way this is correct. However, elementary particles are observed with scientific instruments, and the definition of the object of knowledge concerns also science. The table as an everyday object and the table as a swarm of theoretical objects are the same thing. There is only one table observed from two different physical perspectives determined by sense organs, on the one hand, and scientific instruments, on the other. The pragmatist definition of the object of knowledge holds for both of them. The situation is the same as in the case of a patient seeing a doctor. The same table is interpreted with two different sets of meanings, namely the meanings of everyday discourse and the meanings applied in science.

References

Brentano, F. (1924). *Psychologie vom empirischen Standpunkt, Erster Band*. Leipzig: Felix Meiner.

Dewey, J. (LW 1). (1981). *Experience and nature, the later works 1*. In J. A. Boydston (Ed.), Carbondale and Edwardsville: Southern Illinois University press.

Dewey, J. (LW 4). (1984). *The quest for certainty, the later works 4*. In J. A. Boydston (Ed.), Carbondale and Edwardsville: Southern Illinois University press, 1984.

Hobbes, T. (1962). *Concerning body. The english works*, trl. In M. William (Ed.), (Vol. I). London: Scientia Aalen.

Lakoff, G. & Johnson, M. (1999). *Philosophy in the flesh*. New York: Basic Books.

Locke, J. (1959). *An essay concerning human understanding*. New York: Dower Publications.

Määttänen, P. (1993). *Action and experience. A naturalistic approach to cognition*. Helsinki: Annales Academiae Scientiarum Fennicae B 64.
Noë, A. (2004). *Action in perception*. Cambridge: The MIT Press.
Putnam, H. (1981). *Reason, truth and history*. Cambridge: Cambridge University Press.
Rorty, R. (1980). *Philosophy and the mirror of nature*. Oxford: Blackwell.
Spinoza, B. (1955). *On the improvement of the understanding, the ethics, correspondence*. Trl. In: R. H. M. Elwes (Eds.), New York: Dover Publications.

Chapter 3
Habit of Action

In colloquial language habit of action may refer to blind routine behaviour or bad bodily habits that one should get rid of. In pragmatism it is a central notion in challenging classical philosophy, and it is significant not only in philosophy but also in social theory (see Kilpinen 2000). Even in its pragmatist sense habitual action may be realized automatically, and habits function often subconsciously (CP 2.148, 5.417). This is quite natural for example in the development of skills. A beginning piano player has to think carefully where to put the fingers, but a virtuoso cannot think about such things during the performance. And it is nowadays an established fact that cognition proceeds largely unconsciously (Lakoff and Johnson 1999; Franks 2010).

The real significance of the notion of habit is in its role in dissolving certain false presumptions of classical philosophy. Habits of action are beliefs and meanings for Peirce. To define habits as beliefs is a way to criticize the Cartesian view of beliefs as immaterial ideas. Habits are modes or forms of action performed by biological organisms. Action requires necessarily an environment, and as habits are modes of interaction between organisms and their environment, they cannot be literally internal. The classical dichotomy of internal and external is hereby questioned.

For Peirce habits are also "real generals" or universals. Richard Rorty is right in maintaining that philosophy of mind originated as an attempt to understand how mind grasps universals (Rorty 1980, pp. 38–45). Rorty's goal is to get rid of philosophy of mind in the classical sense, which is typical for pragmatism. He does it by rejecting universals altogether. No universals, no mind in the classical sense and, therefore, no need to a specific philosophy of mind. However, this is not necessary. Habits of action are universals whose mode of existence is, however, somewhat peculiar (Sect. 3.2). If habits are real general entities and are actualised through bodily action, then they form the "mind-stuff" that Rorty writes about (Rorty 1980, p. 30). This stuff is best understood as forms or modes of organism

P. Määttänen, *Mind in Action*, Studies in Applied Philosophy,
Epistemology and Rational Ethics 18, DOI 10.1007/978-3-319-17623-9_3

environment interaction, which is why it cannot be internal. Strictly speaking habits are not properties of the brain or the body (Määttänen 2010).

It is important to note that habit of action is a teleological notion (see Chap. 5). Teleology is a tricky concept for naturalism so far as it tries to stick to hard natural sciences. Physical nature allows of causal processes only. Future cannot have an effect on the past. However, as Peirce pointed out, future can have an effect on the present without affecting the past. Habits are actualised in the loop of perception and action. This loop is realized through physical causal processes, but its properties as a loop explain how anticipation of the future is possible on the ground of past practical experience without violating the principle of the causal closure.

3.1 Habit: A Structured Scheme of Action

Richard Rorty maintains that habits are bodily states "attributed to organisms of a certain complexity" (Rorty 1991, p. 93). Can habits be bodily states? Peirce compared habits with dispositions (CP 5.440), and dispositions are sometimes understood to be properties of individuals. A person is said to have a disposition to act in a certain way in certain circumstances. Habits are, however, better understood as forms of interaction rather than as bodily states. A disposition to act is a relational concept in the sense that there is no disposition to act without (potential) action, and no action without some circumstances. A disposition to act requires a situation and the specific circumstances, which make this action possible. The definition of a disposition refers to these circumstances, and it remains an open question how a relation that consists of a living agent, action and specific circumstances can be considered as a bodily state. Generally speaking, it would be a logical category error to reduce a relation to one of its elements, and habits as forms of interaction are relations between living organisms and their environment. Peirce actually appealed to the role of circumstances when he explained how habits differ from dispositions:

> Habits differ from dispositions in having been acquired as consequences of the principle, virtually well-known even to those whose powers of reflexion are insufficient to its formulation, that multiple reiterated behaviour of the same kind, under similar combinations of percepts and fancies, produces a tendency – the habit – actually to behave in a similar way under similar circumstances in the future (CP 5.487).

The formation of a habit depends on the acting agent and on the circumstances to which action is accommodated. The role of the circumstances is neglected if one considers habits as bodily states. Habits are actualised through the interaction between biological bodies and their environment.

Habits are formed when action is accommodated to objective conditions of action. In habit formation action gets structured according to these conditions. Jean Piaget called these structures sensorimotor schemes. His characterization of the schemes is similar to Peirce's expression above. In Piagetian theory a sensorimotor scheme is an organized series of motor acts, which is formed by reiterating action in the same or similar circumstances (Piaget and Inhelder 1969, p. 4). Piaget borrowed the concept of scheme from Immanuel Kant's doctrine of schematism. After Piaget the term has been used in cognitive psychology and computer science (see Määttänen 1993). The role of motor action and embodiment is also emphasized in George Lakoff's and Mark Johnson's *Philosophy in the Flesh* (1999).

Peircean pragmatism has similar objectives. The central notion is habit of action, which is a structured series of acts reiterated in similar circumstances. Habitual action requires relatively stable conditions: an acting agent and a stable environment. Habits of action get accommodated to "laws or habitudes of nature", to use Peirce's expression (CP 5.587). Peirce discussed these conditions in terms of the so-called uniformity of nature. He did not want to appeal to the uniformity of nature in showing mathematically the validity of induction. However, he says that "a certain degree of special uniformity is requisite" for the possibility of inductive reasoning (CP 2.775). Here we must take notice of Peirce's conception of induction. Habit formation is for Peirce a form of inductive reasoning that is performed on the ground of practical experience (CP 6. 145). Logic is here conceived as logic of action. Logical necessity is based on generality, as was pointed out in Chap. 1, and habits as "real generals" (see the next section) thus belong to the toolbox of logic. To form a habit is to make a general logical conclusion from practical experience. Peirce characterized habit as "real and living logical conclusion" (CP 5.491).

A certain degree of special uniformity is, indeed, required for habit formation. There must be a relatively stable physical environment where acting organisms can live, perceive and act. Action involves effort that meets resistance. This resistance forms the objective conditions of action. The everyday world just has a certain structure of middle-sized rigid objects, one's own body included. This is the basic requirement of any known form of consciousness. There is not much evidence of disembodied mind, after all. The required degree of special uniformity is just this stable environment of living creatures. This precondition of experience is in no way transcendental. It does not depend on any supernatural things. It is a simple empirical commonplace that we experience stability in daily practices.

3.2 Habits as General Entities

Habits are actualised as activities of embodied agents in relatively stable conditions of action. Meanings and beliefs are supposed to be general entities if they are to fulfil their function as vehicles of cognition. In what way are habits general? What is their mode of existence that allows of generality? Peirce approached this question by asking *when* do habits exist? There are three obvious possibilities, past, present and future. Peirce writes:

> For every habit has, or is, a general law. Whatever is truly general refers to the indefinite future; for the past contains only a certain collection of such cases that have occurred. The past is actual fact. But a general (fact) cannot be fully realized. It is a potentiality; and its mode of being is *esse in futuro*. The future is potential, not actual. What particularly distinguishes a general belief, or opinion, such as is an inferential conclusion, from other habits, is that it is active in the imagination. (CP 2.148.)

In the past there has been only a certain finite number of instances of any habit. No genuine generality can be involved here. The same holds for the present because acts are performed one at a time. There remains only the future. How is generality involved here? Peirce writes that general refers to the indefinite future, not infinite future. Further, Peirce refers to potentiality. Compare this with the notions of actual and potential infinity in mathematics. Actually infinite sets, for example the set of natural numbers, are considered to be objects of mathematical operations as a whole, with all the infinitely many numbers in it. A potentially infinite set is a set of natural numbers from zero to **n** where **n** is arbitrarily big. The set is not actually infinite because it has a finite number of members, but it is potentially infinite in the sense that one can always take a bigger number as the upper limit. It is an indefinitely big set.

Analogically actual generality refers to the infinite future, which cannot belong to the sphere of experience of any earthly creature. Life is finite. Immanuel Kant criticized earlier rationalists for making this error of transcending over the limits of experience. As pointed out earlier, Peirce agrees with this and characterizes Kant as a pragmatist, although a confused one. The thing in itself is supposed lie completely outside the scope of experience by definition. But quite consistently with the restriction concerning the use of concepts Kant writes, when discussing the antinomies of pure reason, that questions about time having a beginning and space having a border cannot be solved within the limits of experience and are thus not solvable by human understanding. Actually infinite cannot be accessed in experience.

Habits as meanings can only be applied to experience, and they are only potentially general. They make it possible to anticipate consequences of habitual action into indefinite future. Future cannot be fully actualised, and this entails that habits are doomed to stay potentialities. This, in its turn, entails that habits as meanings and beliefs can only be objects of thought. Strictly speaking habits can never be perceived. One can perceive instances of habitual action, but habituality, the idea that acts are instances of habits, is added by cognition, which in the present analysis is basically anticipation of action. Note that this definition covers also linguistic behaviour (see Chap. 4). Generality in cognition is based on the similarity of action in similar circumstances. Kant's pragmatist character is seen also here, since according to Kant generality is continuous activity (stetige Handlung; Kant 1926, p. 615).

It is important to note also what does *not* follow from this view of habits as general entities. Habits are potentialities that can only be objects of thought. Thus they are cannot be reduced to bodily states or defined in anatomical terms. But these objects of thought do not reside in a specific mental substance. Habits are not independent of our bodily existence. Habits are actualised as individual acts, as physical processes in the physical world. The fundamental error of classical philosophy is the doctrine that entities, which are real and can only be objects of thought, are given an independent ontological status as if they could exist by themselves, think by themselves and be seriously sceptical about the existence of the material world.

Peircean view about habits helps to avoid also another erroneous conception of classical philosophy. Franz Brentano tried to define the subject matter of psychology with an analogy between words and mental states (Brentano 1924, pp. 124–125). According to him mental states are intentional units referring to (real or intentional) objects just like words in a language do. Mental states and words are *about* something else, from which we get the characterization of intentionality as *aboutness*. This view is sometimes naturalized by identifying mental states with brain states (or processes). This move retains the idea that beliefs (vehicles of thought) are individual units located in the brain. The Cartesian conception of mind as a container of ideas is materialized (Lakoff and Johnson 1999, p. 395).

Habits, however, are actualised in the interaction between organisms and their environment, and this interaction consists of a complex system of relations. Habits as beliefs and meanings are not individual units but relations. Neural events are obviously important elements of this system of relations, but as pointed out earlier, it would be a logical category error to reduce a relation into one of its elements. The relation would cease to exist. And it does not make sense to ask for a location of a relation in the same sense, as it is sensible to ask for a location of its elements. This is just not a good question. Beliefs, meanings, thoughts, ideas and so on are relations that strictly speaking don't have a location in the same sense as individual units. At most one can say that the system of relations through which habits are actualised is external to the body, because the interaction with the environment is necessarily involved.

Habits are real in the sense that they have a real effect on how we behave, and their somewhat peculiar mode of existence is the anticipated potential future. They are general in the sense that similar behaviour is repeated in similar circumstances, and the "laws or habitudes of nature" are general in the same sense. The relatively stable and general features of action are accommodated to the relatively stable and general features of environment that form the objective conditions of action.

3.3 Habits as Beliefs

Beliefs are thoughts about the world. As pointed out before, habits as general entities are never fully actualised. They exist as thoughts about the indefinite future. Thinking with habits is essentially anticipation of possible future courses of events. In what sense are habits thoughts about the world?

Habits are schematically structured sequences of acts in the world. Habits are formed during actual behaviour when action accommodates to the objective conditions of action. The structure of habitual action fits with the structure of the world. The schematic structure of a habit is not propositional because language is not necessarily involved. To think with habits is to think about what to do in the world in order to achieve one's goals. The relation between habits and the world is the operational fit between them, and the criterion of the fit is success. If the success is an outcome of anticipated fit, then the habits involved are correct ones; they supply true knowledge about what to do. Of course, one can achieve goals accidentally, but anyway the activity has proceeded in prevailing objective conditions to which behaviour is accommodated. This is how new and perhaps better habits (beliefs) are formed.

Habit formation is induction. In habit formation one acquires general conclusions on the ground of past practical experience. The world has general features (laws or habitudes of nature, and there are regularities in the social world as well). In habit formation the habitual behaviour gets general features during the practical experience when action accommodates to the general features (regularities) of the world. In this sense one can experience generality, although one cannot strictly speaking perceive generality, as the classics of empiricism correctly pointed out. One can perceive only single acts when they are performed. The acknowledgment of the fact that generality can be experienced—and is not just something that is added to experience by reason separated from experience—is an important difference in comparison with empiricism. In pragmatism reason is not separated from experience. On the contrary, practical experience, habit formation, is a mode of reasoning that helps one to anticipate, that is, to reason, about the possible future courses of events. Reasoning with habits is the basic way to experience the world, and habit formation is a way to experience generality embedded in the world as "laws or habitudes of nature" as Peirce said. Also the use of language proceeds within experience. Not a word is emitted without bodily behaviour, and symbols gain meaning when they are used in the context of other practices.

Habits as beliefs are vehicles of cognition, but how do we think with habits? Peirce describes the issue with the following illustration.

> In a piece of music there are the separate notes, and there is the air. A single tone may be prolonged for an hour or a day, and it exists as perfectly in each second of that time as in the whole taken together; so that, as long as it is sounding, it might be present to a sense from which everything in the past was as completely absent as the future itself. But it is different with the air, the performance of which occupies a certain time, during the portions of which only portions of it are played. It consists in an orderliness in the succession of sounds which strike the ear at different times; and to perceive it there must be some continuity of consciousness which makes the events of a lapse of time present to us. We certainly only perceive the air by hearing the separate notes; yet we cannot be said to directly hear it, for we hear only what is present at the instant, and an orderliness of succession cannot exist in an instant. These two sorts of objects, what we are immediately conscious of and what we are mediately conscious of, are found in all consciousness. Some elements (the sensations) are completely present at every instant so long as they last, while others (like thought) are actions having beginning, middle, and end, and consist in a congruence in the succession of sensations which flow through the mind. They cannot be immediately present to us, but must cover some portion of the past or future. Thought is a thread of melody running through the succession of our sensations. (CP 5.395.)

To think about habitual action is like thinking about a melody. On the ground of the past one anticipates the future course of events. The structure of experience is here different: the hidden causes of perception are replaced by anticipated consequences of action. The object of knowledge is changed accordingly. Causes and effects are changed to means and consequences. To think with habits is to think about the consequences of action, consciously or subconsciously.

Cognition with habits is not a routine procedure. The world is experienced as possibilities of action, and there are always a large number of possible habitual acts and their combinations. These possible courses of action are elements that can be combined in different ways. Habits can be changed during this anticipation. Peirce called these mental acts of anticipations "fancied reiterations" (CP 5.487). An act may well be habitual even if it is a completely new one and perhaps performed only once. This fancied reiteration is rational cognition. The locus of rationality is in habituality, as Erkki Kilpinen puts it (Kilpinen 2000, pp. 68–71). Habits leading to anticipated goals are rational beliefs about the world. For short: "Thinking is anticipation of action, beliefs are habits of action, and also meanings are habits of action" (Määttänen 2009, p. 112).

There is some evidence supporting this view. The so-called mirror neurons may well belong to the neural mechanisms doing the anticipation (Brincker 2012). "It has been proposed that internal motor representations, also known as forward models, serve as predictors in the brain" (Choudhury and Blakemore 2006, p. 40). The idea of forward modeling is applied also in language production and comprehension (Pickering and Garrod 2013). Also Brendan Johns and Michael Jones have presented and tested a model of expectation generation in sentence processing (Johns and Jones 2014). Andy Clark is quite optimistic about this. "Action-oriented predictive processing models come tantalizingly close to overcoming some of the major obstacles blocking previous attempts to ground a unified science of mind, brain, and action" (Clark 2013, p. 200).

That cognition is not routine does not entail that cognition is always conscious. A consideration of the development of any skill, like playing piano, shows that in the beginning one must concentrate consciously to simple tasks until they are learned. Then one can concentrate to other issues. What is subconsciously performed now is more or less consciously learned earlier. And this earlier experience covers the whole of our biological and cultural evolution.

3.4 The Pragmatist Law of Association

A habit makes it possible to create an association between an observed situation and a future situation, which will appear as a result of habitual behavior. This sort of an association, which can be called the pragmatist law of association, is not included in David Hume's principles of connection among ideas: resemblance, contiguity in time or place and cause (or effect). These classical laws of association are, in a form or another, still effective in contemporary work in artificial intelligence. One example is Teuvo Kohonen's work on self-organizing neural networks (1988, p. 3). Kohonen discusses the phenomenon called autoassociative recall of missing fragments (Kohonen 1988, pp. 160–163). Suppose that a photograph of a human face is stored in an associative memory. When a fragment of the face is used as a key pattern, the network is able to reconstruct the whole face as an output. This is due to the associative connections, which have been formed between the nodes of the network during the storing process. A similar approach can be used for processes that proceed in time. Kohonen describes networks that can store temporal sequences (Kohonen 1988, pp. 16–18). The rest of the stored sequence is recalled by using its first item as a key pattern. The important question is, of course, what gives the order to the sequence. In Kohonen's version of the classical laws of association it is simply the fact that they occur in close succession, that there is a "temporal contact" (Kohonen 1988, p. 3). Items are stored in the memory one after the other.

The pragmatist law of association differs from this in that the associative connections between items are formed not only because they occur in a sequence, but because they are associated with a certain form of action, a habit. Sensory inputs are associated not only with each other but also, and more importantly, with neural mechanisms controlling overt motor action. It is the course of habitual action that determines what kind of sensory inputs are relevant for guiding successful action and therefore associated with one another in a sequence, and what sequence of the neural processes controlling motor movements is associated with it. The important point is that when habitual action determines the sequence of sensory inputs that are associated with each other, the sequence corresponds to the objective conditions of action to which the action is accommodated. Associative process selects properties that facilitate interaction with the world (Rizzolatti and Sinigalia 2008, p. 34; Franks 2010, p. 88). The operational success explains why the habit has become what it is, and it explains also why the sequence of sensory inputs associated with each other is what it is. The operational success is the criterion for picking up the stored items from the temporal flow of sensory input. A mere temporal contact is not enough.

It is important to keep in mind that Hume's associative psychology has a completely different general framework as a background. Sensory qualities like colors, pain, and heat and cold are "perceptions in the mind" (Hume 1978, p. 469). The epistemological problem here is the relation between internal perceptions and external objects. And the relations formed with the laws of association between internal perceptions (or ideas) are even more difficult to handle because relations like causation cannot be literally perceived. Te associations are formed on the ground of the internal qualities of perceptions. For Hume experience remained silent about the real relations like causality between external objects.

Pragmatism puts the knowing subject inside the world, in the midst of a web of different interactions. An acting agent meets the resistance of the objective conditions of action in the world. There obviously are relations between internal neural events in the brain. Contemporary brain imaging methods show that there are correlations between neural events and things in the world (external to the body). When an acting agent is investigated as one element in a system of interactions there is no need to look for literally internal features that would explain the formation of associative chains of neural events. Neural events are not associated to each other by virtue of internal (neural) features but by virtue of their connectedness to overt action.

The pragmatist law of association does not require that neural events are treated as internal intentional units, and neither does it require any internal mechanism for manipulating these internal processes. They get manipulated through practice. Practical experience creates associative chains on the ground of what perceptions turn out to be relevant for the successful behavior. What get associated with each other are not only some internal neural events. They are neural events that are related to an anticipated course of organism environment interaction due to a chain of overt acts in certain circumstances.

These associative chains are then used for anticipating possible outcomes of habitual action in different situations. If the conditions are similar and action proceeds in a similar way, then it is probable that the situation to be encountered as an outcome of anticipated action would also be similar. The associative chain created during habit formation activates (partly) the same neural events that would be activated when and if the situation occurs. How are these chains activated? At the simplest level they are manipulated by moving around in the observed environment. One activates different anticipatory mechanisms simply by looking at different things (recall that in this view the world is experienced as possibilities of action; for empirical evidence see Franks 2010, pp. 85–91). Habits function as vehicles of cognition as elements of the ongoing interaction, and the active agent is the biological organism as a whole. In pragmatism the knowing agent is not just looking at the external world searching for intentional entities capable of referring to something. This holds also for linguistic cognition. The problem of the meaning of words is *not* posed as "What gives the black dots 'table' the capacity to refer to different tables?" but rather: "How are the habitual ways of using the word 'table' related to other habitual activities having something to do with tables?"

Similarly, the problem of mental content is *not* posed as: How is some mental content related to some unit in the brain and/or to things in the environment? But rather: What is the role of brain states and processes in controlling human behavior, especially in using language and other symbolic systems? The idea that habits of action are vehicles of cognition is an alternative to views based on the Cartesian dichotomy of external and internal and on the assumption that there are internal units representing the external world. The basic claim of this alternative is that cognition requires interaction with our natural and cultural environment and that habit of action is one of the key concepts in analyzing this interaction. Internal mechanisms have a role in controlling behavior but the brain as such is not a subject that knows, thinks and wills. The organism as a whole does all this.

The pragmatist law of association gives also a simple answer to the traditional problem of what gives the structure and order to experience. The traditional Kantian answer to this question is that the pure forms of sensibility and the pure concepts and categories of understanding do the job. This is often expressed by saying that conceptual framework carves up the world. Unfortunately there are not too many explications of what is the concept of concept appealed to in this context. Richard Rorty's appeal to language does not address the problem of from where comes the order in language and how did we get the ability to experience order in language (see Määttänen 2006). If one cannot see order in nature, in trees rocks and animals, how can one see order in verbal units? There is evidence for the opposite view. "There is now a body of evidence for an embodied view of language, according to which language comprehension is based in our bodily experience of the world and involves the same systems necessary for bodily experience" (Perniss, Vinson, Fox and Vigliocco 2013, p. 1133; see also Knott 2012).

Two basic assumptions of classical philosophy, namely the dichotomy of apparent and real and the separation of reason from experience, tempt one to think that the order of everyday world is something unreal. As noted before, pre-Socratic philosophers had difficulties in conceptualizing movement and change, and the result was the Platonic division of the world into two levels, moving and somehow less real empirical world and real world of ideas (or forms). Ideas are supposed to be eternal and unchanging which alleged fact supported the classical ideal of true knowledge: timeless and unchanging conceptual truths achieved by reason independently of empirical experience. And if one wants know something about the empirical world, then our experience of it has to be stabilized with ideas, words or concepts.

These classical presumptions are not valid. There is only one world wherein biological organisms like us live. Naturalism does not allow of human reason capable of having true knowledge of any supernatural spheres that cannot in principle be accessed by empirical methods of inquiry. This is not to underestimate the role of abstract conceptual analysis. But there just have to be some connection to experience if this analysis is to be of some use in acquiring knowledge about the world.

The present bottom-up strategy of explaining how experience gets structured starts from the simple condition that we are biological creatures living in nature. The first and absolutely necessary condition of experience is our existence as embodied beings. There is no evidence of any disembodied experience. Acting organisms must accommodate to objective conditions of action. These physical conditions of experience are in no way transcendental. These conditions are experienced all the time in everyday practices.

The basic conditions of action are experienced by muscular effort and resistance, as Peirce puts it in several occasions. "The environment becomes objectified only in relation to the animal's *motor* capacities" (Franks 2010, p. 88; emphasis in the original). The everyday world has a structure of three-dimensional rigid bodies, some of which are living creatures. When organisms move in this environment the muscular effort meets physical resistance. And when action gets accommodated to this structure, then the structure is imprinted into the schematic structure of habitual action. This imprinting is possible even without the sensory activities of biological bodies (Määttänen 1993, pp. 64–65, 1997). The spatial structure of the visual field is not an outcome of conceptual interpretation and rational organization coming from separate reason, from somewhere above experience. The rigidity of the physical environment is experienced as soon as we take action into the notion of experience. The pragmatist law of association gives the ground for organizing perceptions in a way that enables one to fit one's beliefs (that is, habits) to the structure of the world (objective conditions of action).

So the question was: Why is the world perceived as having an order and structure and not as chaos of perceived qualities, as perceptual manifold? The traditional answer appeals to ideas, language or concepts, to some capacity of reason that is exerted upon what is perceived. The pragmatic answer is radically different. The world is structured as it is, and the bodies of living organisms form one element of this structure. These organisms simply have to accommodate to objective conditions of action. During habit formation the pragmatist law of association connects those perceived features of environment that are relevant for successful accommodation. The structure of the world is experienced in action and imprinted in the structure of habits, which are involved also in perception.

References

Brentano, F. (1924). *Psychologie vom empirischen Standpunkt*. Leipzig, Felix Meiner: Erster Band.

Brincker, M. (2012). If the motor system is no mirror. In N. Payette (Ed.) *Connected minds: cognition and interaction in the social world* (pp. 158–182). Newcastle: Cambridge Scholars Publishing.

Choudhury, S., & Blakemore, S.-J. (2006). Intentions, actions, and the self. In S. Pockett, W. Banks & S. Gallagher (Eds.), *Does consciousness cause behavior?* (pp. 39–51). Cambridge & London: MIT Press.

Clark, A. (2013). Whatever next? Predictive brains, situated agents, and the future of cognitive science. *Behavioral and Brain Sciences, 3*(36), 181–204.

Franks, D. (2010). *Neurosociology. The nexus between neuroscience and social psychology*. New York: Springer.

Hume, D. (1978) *A treatise of human nature*. In L. A. Selby-Bigge (Ed.). Oxford: Oxford University Press.

Johns, B., & Jones, M. (2014). Generating structure from experience. In *Proceedings of CogSci 2014 in Quebec*, pp. 254–259.

Kant, I. (1926). *Reflexionen zur Metaphysik. Kant's gesammelte Schriften 17*. Berlin & Leipzig: Preussische Akademie der Wissenschaften.

Kilpinen, E. (2000). *The enormous fly-wheel of society. Pragmatism's habitual conception of action and social theory*. Helsinki: Department of Sociology, University of Helsinki (Research Reports No. 235).

Knott, A. (2012). *Sensorimotor cognition and natural language syntax*. Cambridge: The MIT Press.

Kohonen, T. (1988). *Self-organization and associative memory* (2nd ed.). Berlin: Springer.

Lakoff, G., & Johnson, M. (1999). *Philosophy in the flesh*. New York: Basic Books.

Määttänen, P. (1993). Action and experience. A naturalistic approach to cognition. Annales Academiae Scientiarum Fennicae B 64, Helsinki.

Määttänen, P. (1997) Intelligence, agency and interaction. In G. Grahne (Ed.), *SCAI'97, Sixth Scandinavian Conference on Artificial Intelligence* (pp. 52–58). Amsterdam: IOS Press.

Määttänen, P. (2009) *Toiminta ja kokemus. Pragmatistista terveen järjen filosofiaa* (Action and Experience. Pragmatic philosophy of common sense). Helsinki: Gaudeamus.

Määttänen, P. (2010). Habits as vehicles of cognition. In A.-V. Pietarinen et al. (Eds.), *Applying Peirce. Proceedings of International Peirce Conference, Helsinki, June 2007*. Helsinki: Nordic Pragmatism Network. http://www.nordprag.org/nsp/1/.

Perniss, P., Vinson, D., Fox, N., & Vigliocco, G. (2013). Comprehending with the body: action compatibility in sign language? In *Proceedings of CogSci 2013 in Berlin* (pp. 1133–1138).

Piaget, J., & Inhelder, B. (1969) *The psychology of the child*. London: Routledge & Kegan Paul.

Pickering, M., & Garrod, S. (2013). An integrated theory of language production and comprehension. *Behavioral and Brain Sciences, 4*(36), 329–347.

Rizzolatti, G., & Sinigalia, C. (2008) *Mirrors in the brain: how our minds share actions and emotions* (F. Anderson, Trans.). Oxford: Oxford University Press.

Rorty, R. (1980). *Philosophy and the mirror of nature*. Oxford: Blackwell.

Rorty, R. (1991). *Objectivity, relativism, and truth. Philosophical Papers* (Vol. 1). Cambridge: Cambridge University Press.

Chapter 4
Habits as Meanings

According to Peirce, what a thing means is simply what habits it involves (CP 5.400). Dewey gave essentially the same definition by saying that action and its consequence must be joined in perception, which relation gives meaning (Dewey LW 10 1987, p. 51). This definition can be applied to symbols as well as to other objects of perception. The outcome of this analysis is a layered system of meanings with which we experience, interpret and understand our environment.

4.1 Meaning: A Three-Place Relation

The minimum requirement of meaning is that with meanings one can think about something that is not here and now but somewhere else at another time. Meanings give cognitive distance in regard to the immediately given situation. An observed meaningful entity leads thought to something else. This obviously requires three things. There has to be someone who observes, that what is observed (sign-vehicle) and that towards which thinking is guided. The sign-vehicle brings to mind its object (that what is thought with the sign-vehicle). Sign-relation thus has necessarily three components. It is a three-place relation. Iconic signs refer to their objects by virtue of their own properties, for example similarity. But similarity alone is not enough for a meaning relation. The iconic relation must become interpreted as a sign-vehicle referring to some similar thing. Three components are necessarily needed because a sign-vehicle's capacity to refer must be given an explanation. A couple of black dots on a paper cannot have any magical property of referring just by themselves. It is not the capacity of these dots to refer. It is the reader's capacity to understand that these dots form a meaningful unit and refer to something else, that they have a meaning. A third element is always required.

P. Määttänen, *Mind in Action*, Studies in Applied Philosophy,
Epistemology and Rational Ethics 18, DOI 10.1007/978-3-319-17623-9_4

Analysis of meanings is sometimes insufficient in this respect. According to David Papineau naturalist theories of meaning appeal to mental contents (Papineau 2006, p. 175). What is the mode of existence of these mental contents? Apparently a Cartesian mental substance would not be the choice for a naturalist. One alternative is to identify mind with the brain. In this case one should explain what exactly gives a neural event the capacity to refer to something. By virtue of what is it an intentional unit, literally internal representation? As we shall see in the next chapter, the claim that mental events are realized through brain processes has its problems. Here it suffices to say that the appeal to mental contents in the brain only answers the question (How do words refer?) with another question (How do neural events refer?). Mental contents are the topic of the next chapter, but obviously one cannot apply the strategy in the second question. Explaining mental contents with other mental contents is flatly circular.

Both questions make it explicit that the capacity of a word or another sign-vehicle to refer needs an explanation. A two-place relation between a sign-vehicle and its object is not, as such, a meaningful relation. There is must be some real persons to understand that the word refers. The sign-vehicle must be interpreted to refer to its object.

4.2 Habits as Tacit (Non-linguistic) Meanings

Some pragmatists take meanings as a thoroughly linguistic issue. One example is Richard Rorty who has been characterized as a linguistic turn pragmatist (Hildebrand 2003). Rorty's claim that language gives the structure and order to nature is one manifestation of it. This narrow notion of meaning excludes the possibility of non-linguistic meanings. Peirce, however, is definitely not a proponent of such narrow notion of meaning. Habits are not exclusively linguistic activities. Peirce's wide notion of meaning shows the possibility of non-linguistic habits that are meanings of their own right, independently of any contribution of symbolic languages.

The notion of tacit knowledge has become famous from Michael Polanyi's writings. Polanyi introduces the notion as an alternative to knowledge expressed in language. According to Polanyi (1969) skills and sense perception are special forms of knowing that can be called tacit knowledge. He mentions also meanings in this context but is not too explicit about what notion of meaning he applies. Peirce's definition of meaning as habit of action serves well in the case of skilful behaviour. The idea is not to subscribe to everything that Polanyi writes about tacit knowledge. There just seems to be certain similarities and connections between the Peircean notion of meaning and Polanyi's central themes. This is not the proper place to explore this in detail. Non-linguistic habitual practices are meaningful and deserve to be called tacit meanings. These tacit meanings are vehicles of cognition, and the stable results of tacit thought deserve to be called tacit knowledge. This is perfectly consistent with the earlier definition of the object of experience and knowledge as a relation between two situations mediated by habitual action.

The pragmatist definition of meaning entails the idea that one can anticipate the consequences of action in a given situation. This gives meaning to the situation in the sense that different observed aspects and things (which involve habits) refer to these potential consequences. One understands what will probably be experienced if some habits or their combinations are performed. This fulfils the minimum requirement of anything's having meaning. Habits give cognitive distance to the immediately observed situation.

Virtually anything observed may involve habits and, therefore, be a meaningful sign-vehicle. The only thing required is that it brings to mind, consciously or subconsciously, some habitual behaviour with anticipated consequences. This means that the experienced world is always interpreted with meanings. It is important to note here that language is not necessarily involved in this interpretation. In the case of human beings language no doubt has an effect on this, but in principle this system of tacit meanings is a vehicle of meaningful cognition of its own right.

Anticipation is based on experience, which is a complex notion. Evolutionary experience has shaped our biological structure. The ability to perceive certain qualities, for example, is based on evolution. Sense organs are kind of crystallized habits of picking up those features of the environment that have been relevant for survival. Similarly there are many inborn mechanisms of instinctive habitual activity participating in the more or less conscious planning of behaviour. In a similar manner we have a long history of cultural evolution. Every individual must be socialized into some human community, learn to speak and get educated to some extent in order to realize her human potentiality. Individual experience determines what one eventually adopts and learns from the social environment. Experience is a complex and layered background that helps to anticipate possible consequences of habitual action. Most of this takes place subconsciously because the vast majority of the actual experiences that have shaped our storage of habitual facilities during our natural and cultural evolution cannot be consciously accessed. This can be called the new unconscious as distinct from the views of Sigmund Freud and others (Franks 2010).

One notable feature of tacit meanings is that they are not particularly conventional. This is due to the fact that typical sign-vehicles are physical objects like tables and chairs, doors and windows. Most of us have the habit of using a door and not a window when exiting a room. This entails that doors and windows have a different meaning for us. One can easily anticipate what kind of experiences will be encountered in each case. This difference in meaning is dictated by objective conditions of action, the physics of rigid three-dimensional objects in this case. In the case of tools and instruments we can apply the principle that meaning is use. A physical tool can be used for different purposes, but the possible ways of using the tool are restricted by the physical properties of the tool as well as by the properties and powers of the one who uses the tool. In symbolic language the physical properties of the sign-vehicles are utilized only for distinguishing the letters and words from each other. These properties do not restrict their use in any way. But this does not hold for tacit meanings.

Also single perceived qualities are carriers of meanings (Dewey LW 10 1987, p. 122). There are no "pure" or "simple" qualities because they are qualified by implicit reactions of many organs (ibid., pp. 126–127), charged with subconsciously anticipated consequences (largely on the ground of our evolutionary experience). These hidden consequences just form the meanings of single qualities. And because of the fact that these habitual anticipations are part of our inherited structure and instincts they are not so conventional either. Of course, the cultural layer of meanings has an impact on tacit meanings, but tacit meanings are genetically earlier, not so conventional, and in this sense more fundamental.

Dewey explained the character of tacit meanings (although he did not use this term) when he discussed meanings that are typical in art as distinguished from language. Words are symbols that represent, stand for or point to something else (Dewey LW 10 1987, p. 89). Science states but art expresses meanings. "Statement sets forth the conditions under which an experience of an object or situation may be had" (ibid., p. 90). Statement may lead to an experience. But expression "does something different from leading to an experience. It constitutes one" (ibid., p. 91). Tacit meanings of art are present in the picture, embodied or incorporated in it. These meanings are also individualized in the sense that any change in the colours or lines in the picture changes the meaning, while the font or colour of a textual statement may change without affecting the content. Meanings in art are also emotionally powerful (Määttänen 2012, pp. 109–119, 2015). Meanings in art are thus individualized and emotionally expressive, but on the other hand, they must be general in order to be meanings. How is that?

Habits as meanings explain this fairly well. Habits are general entities. On the other hand they are fuzzy and context dependent because circumstances are never exactly similar, and habitual anticipation gives only the schematic structure of the intended activity. Anticipation typically remains approximate. This is why actualisation of any tacit meaning is individualized. Mental habits, that is, mental reiterations of habitual action produce new ways of activity. A picture as well as any scene of action in real life is full of conscious or subconscious tacit meanings, which refer to various consequences of possible habitual activities. Any change in the perceived situation changes some meanings and the possible ways of combining different possibilities of action. Tacit meanings in real life are also closer to the emotional mechanism than abstract word meanings. For Dewey works of art are experiences, and as realized experiences they take similar distance to abstract word meanings as tacit habits. This flowing system of tacit meanings is extremely rich and forms the stream of experience upon which the more exact and definite system of linguistic meanings is built (see Lakoff and Johnson 1999; Donald 2002; Johnson 2007).

4.3 Linguistic Meaning

In Ludwig Wittgenstein's language games linguistic meaning is defined by the principle that meaning is use. The meaning of a word is the way it is used in a linguistic community. It is easy to see that Wittgenstein's definition is an instance of Charles Peirce's definition (Määttänen 2005). Surely the habits of using a word belongs to the habits the word involves. According to John Dewey the word "hat" gains meaning in the same way as hat, by being used in a given way (Dewey MW 9 1980, p. 19). Dewey's definition makes explicit the parallel between words and other things that can be used, which is implied by Peirce's definition. It does not restrict the things (involving habits) in any way. Also Wittgenstein refers to the analogy between words and tools. Is this parallel between words and tools mere analogy?

If it is a mere analogy, then tools and other things involving habit are not carriers of meaning of their own right. Language games would contain only linguistic moves. This raises the problem of the origin of linguistic meanings. Rows of letters have no meanings just by virtue of their physical properties. This is a kind of semiotic circle: words of a language are defined by other words of the same language. How to get from the language to the world?

As pointed out, it is problematic to explain the ability to understand language by appealing to mental contents. Mere postulation of mental entities remains circular without further explications. The problem of how we understand language is solved by appealing to entities that are simply thought into real being, and the only thing we get to know about these entities is that they solve the problem that we had in the first place. Postulation of objective immaterial meanings is not consistent with naturalism and anyway circular in the same way. On the other hand, ostensive definition of words by pointing to something and giving it a label has its difficulties also. For example, if we try to teach somebody a completely new language by pointing to something and uttering "gavagai" (Quine 1960), there is no way of avoiding the problem of aspect. What aspect of the observed thing or situation are we pointing to? Further explanation with the same language will not help. This kind of problems led Wittgenstein to develop his notion of language game with its pragmatist notion of meaning.

The basic error in getting stuck with the problem of aspect concerns the notion of experience. Ostension and sense perception do not give solution. The pragmatist notion of experience, where action is included, and the wide notion of meaning, where also tacit (non-linguistic) meanings are included, give the way out. The parallel between words and tools is not a mere analogy. Tools and instruments, doors and windows and indeed anything that may involve habits are possible carriers of meaning. Meaning is defined in the same way at both levels of meaning (tacit and linguistic), namely by saying that meanings are habits of action.

As pointed out, the traditional way of putting the question is also misleading. The problem is not how a row of letters gets its ability to refer to something else, its intentionality (aboutness). It is the interpreter who has the ability to understand the reference. Natural language gains meaning by being used in the context of other meaningful practices. This implies that linguistic meaning depends on the context. Literal meaning and analysis based on giving the necessary and sufficient conditions (or exact borders) cannot be *the* universal criterion of rational conceptual analysis. The applicability of this kind of criteria depends on the topic. Natural sciences allow of exact definitions simply because some parts of nature are well categorized. The social world is different. Things are overlapping and borders fuzzy. This is probably the reason why Dewey criticized all forms of compartmentalization in his philosophy of art.

Merlin Donald emphasizes that language "floats on the sea of metaphor" (Donald 2002, p. 282). He refers to the work of Mark Johnson and George Lakoff who maintain that reason is "largely metaphorical and imaginative" (Lakoff and Johnson 1999, p. 4). In their view metaphorical thought is based on sensorimotor experience. Much of conceptual inference is sensorimotor inference (ibid., p. 20). They use often the notion of motor schema. This is interesting because, as already pointed out, Jean Piaget's definition of sensorimotor schema (Lakoff and Johnson do not refer to Piaget, but the connection is obvious) and Peirce's definition of habit of action are almost word by word similar (Määttänen 1993, pp. 55, 59n).

The idea of the present approach is the attempt to analyse sensorimotor experience and metaphorical thought with an explicit notion of tacit meaning: habits of action (or sensorimotor schemas) just are meaning relations with which meanings may be associated to any object of perception. An object of perception is a sign-vehicle because it refers to the previous and anticipated consequences of habitual action the object in question involves. Linguistic meaning evolves on the top of tacit (non-linguistic) meanings. The relation between natural language and metaphorical thought is a relation between two different systems of meaning.

4.4 The Layered System of Meanings

Meanings are habits of action. This unifying principle of the system of meanings explains the emergence of natural language, ties linguistic meanings with experiential practices and restricts the conventionality of linguistic meaning.

A traditional way to explain the emergence of natural language is to appeal to the emergence of tools as first consciously manipulated sign-vehicles (see Määttänen 1993). Tools can be used for communication about the things to which tools refer: the objects of tool-use and the activities with which tools are used. Tools are carriers of meaning independently of language, and "words grow to the meanings" that already exist, as Martin Heidegger puts it (Heidegger 1986, p. 161). The ability to take words as carriers of meaning is based on the existing ability to take tools as meaningful things. The meaning of tools is in the habitual ways of using the tools.

Donald's (2002) approach is different. He appeals to early mimetic culture capable of communicating conventional symbols before the emergence of natural languages. He does not even mention tools. According to Donald human mind is a hybrid. It is a product of brain-culture symbiosis. "We are what we are because of enculturation" (Donald 2002, p. 151). The problem is that contemporary culture cannot function without languages, but brains cannot generate languages without pre-existing symbolic cultures (see also Franks 2010, Chap. 3). The solution is that expressive culture must have taken the first step (Donald 2002, p. 250). This is a culture of mimetic expressive skills, which started the development of a genuinely socially distributed cognitive system. In mimetic culture conventional meanings were communicated with mime, imitation, skill, and gesture (ibid., p. 263) As a cultural product language has its mimetic roots. "In their creative origins, symbols are a product of thought, not vice versa, and in their interpretation, symbols get their meaning from thought, not vice versa" (ibid., p. 276).

According to Merlin Donald the brain is an analogue device, and such devices "do not employ symbols in the classic definition of that term" (Donald 2002, p. 102). The machinery of language is not coded entirely inside the genome of a developing brain. "It was more probably shaped by the demands of a communicative universe that was much larger than one contained inside a single brain and was instead provided by a community of brains" (ibid., p. 253). Human mind is a distributed cognitive system. Symbolic minds are "hybrid products of a brain-culture symbiosis" (ibid., p. 202). Franks says the same referring to the work of Leslie Brothers. Human brain needs the presence of other human brains (Franks 2010, p. 39). It takes numerous brains to make one brain work, and these brains do not work without language (ibid., p. 42).

However, after saying that analogue brain does not employ symbols "in the classic definition of that terms" Donald goes on by claiming that symbols "can be internal or external to the brain" (Donald 2002, p. 305). He is not quite explicit what he means by the "classic definition" of symbol. If we accept the idea that symbols and other representations are three-place relations where sign-vehicles refer to their objects only if they are interpreted to do so, then we have a serious problem of how internal representations become interpreted. Nobody has access to one's own brain states or processes. There are no homunculi in the brain that might do the job. As will be argued in Sect. 5.1, it is better to follow Bennett and Hacker and reject all attempts to talk about internal representations. The claim that we have access to internal representations is "no less mysterious than the Cartesian claim that the mind has access to an image on the pineal gland" (Bennett and Hacker 2003, p. 147).

The human brain is genuinely social and requires language as a social phenomenon. Which one came first? Merlin Donald's solution to this chicken-egg problem is mimetic culture that enables communication of conventional meanings without the aid of language. By mimesis "we can share tacit knowledge" (Donald 2002, p. 266). Mimesis is logically prior to language because without it, "we cannot rehearse or refine any skill, let alone one as complex as speech or language" (Donald 2002, p. 268). Mimetic skills make speech possible.

Donald's position raises a problem that he does not address. In order to speak about meanings and knowledge prior to and independently of language one needs a notion of meaning that is prior to and independent of language. Donald does not give one, but his characterizations point to the direction of the pragmatist notion of meaning presented here. The development of mimetic skills requires, according to Donald, that "the focus of attention is not the reward or punishment that follows an act, or its social consequences, but the *form* of the act itself" (ibid., p. 272; emphasis in the original). Donald is here talking about Peirce's deliberately formed and self-analysing habits (see CP 5.491). Donald writes also that the cognitive core of mimesis is "kinematic imagination" (Donald 2002, p. 271). A mimetic controller brings "a variety of action systems under unified command" (ibid., p. 269). It is a kind of virtual space where the actor can deliberately review and modify action in imagination.

Kinematic imagination, conscious review and rehearsal, is central to mimetic action. The mimetic controller is a neural system controlling an infinity of possible forms action involving virtually any muscle group (ibid., pp. 268–272). This can be compared with what Peirce writes about habits.

> A cerebral habit of the highest kind, which will determine what we do in fancy as well as what we do in action, is called a belief. The representation to ourselves that we have a specified habit of this kind is called a judgment. A belief-habit in its development begins by being vague, special, and meagre; it becomes more precise, general, and full, without limit. The process of this development, so far as it takes place in the imagination, is called thought (CP 3.160).

And "a deliberate, or self-controlled, habit is precisely a belief" (CP5.480). Habits as meanings and beliefs form precisely the cognitive machinery for Donald's mimetic culture. This explains a little more explicitly how mimetic culture "gives us a semantic base, a means of referring language outside itself" (Donald 2002, p. 280).

In other words, the Peircean definition of habits as meanings is a way to describe how mimetic cognition actually functions. Mimetic action refers to actual action (in the past or in the future) by virtue of iconicity once the form of action (habitual, schematic structure of action) has become an object of conscious thought. Although Donald does not mention tools and instruments, this definition of meaning brings also this aspect in the big picture. Tool-use and mimesis are not mutually exclusive explanations of the emergence of human cognition. Habits as meanings function as a unifying principle also here.

The layered system of meanings ties linguistic meanings to practical experience. It is, of course, possible to entertain meaning contents that have no obvious connection to tacit meanings. Natural language is a powerful vehicle, and it is perfectly possible to make fiction and poetry, but as far the goal is to discuss the way the world is, it is advisable to keep the system of meanings coherent. Tacit meanings are essentially embodied. Biological organisms act in a physical environment of middle-sized three-dimensional objects. Typical sign-vehicles like tools, chairs and so on are such objects. Typical objects of these sign-vehicles are also such objects. This is how things are quite independently of what scientists find out about the ultimate constituents of the physical universe. Everyday 3d-objects consist of these constituents anyway.

The emphasis of the connection of natural language and tacit meanings is, in effect, a development of Peirce's pragmatic maxim that appeals to practical bearings. Conceptions, distinctions and problems that have no practical bearings are not worth considering in the discussion of what the world is like and what is our place in it. Tacit experiential meanings are the basic ones, and language gains it meanings only in this context.

The connection between language and tacit meanings affects also the conventionality of linguistic meanings. The physical features of letters do not restrict the use of words in any way, but tacit meanings are not so conventional. The use (and thus meanings) of typical non-linguistic sign-vehicles is restricted by hard facts like muscular effort and resistance (see, for example, CP 5.45). The system of meanings should be consistent, and this requirement restricts also the conventionality of linguistic meanings. Experience is also necessarily connected to our existence as embodied beings, and this has an effect on colour concepts, for example. The biological structure of eyes and brains is a basic fact that cannot be ignored.

Tacit meanings are fundamental, but this is not to say that language and other cultural phenomena do not have any effect on tacit meanings. Cognition with meanings is complex activity where things are mutually interdependent. It remains, however, a fact that we are biological organisms who have developed the social and cultural environment around us. Culture is a product of nature, as Dewey put it.

4.5 Experience and Interpretation

The traditional way of explaining how the experienced world gets its structure is the appeal to concepts or meanings. The world is experienced as meaningful by virtue of meanings, which sounds self-evident. It started with Plato who suggested that ideas or words give some stability to the changing empirical world. Immanuel Kant appealed to the pure forms of sensibility (time and space) and to the pure concepts and categories of human understanding. Neo-Kantians criticize the so-called myth of the given and emphasize that experience is always interpreted with concepts, which carve up the world (Sect. 3.4). Any attempt to disagree is labelled naïve empiricism or faulty of the error of admitting the myth of the given.

The fatal error of this view is the conception that concepts or meanings are ultimately linguistic ones and that concepts formulated in language are independent of experience. The pragmatist notion of meaning gives the way out of this dilemma. The system of tacit meanings is prior to and independent of language (although linguistic meanings have an effect on tacit meanings once language has emerged). Tacit meanings (habits of action) are formed on the ground of objective conditions of action faced in daily practices. Knowledge of these conditions is based on practical experience. Mind is embodied and bodily action is accommodated to the rigid environment. No amount of language use can change this. Thus it is rather odd to take seriously the possibility that "gavagai" may refer to a slice of rabbit instead of to a rabbit as a whole. The natural assumption is to start the interpretation with rabbit as a whole, and if some problems in discussion occur, they can probably be solved by referring to some concrete measures, by using a knife and asking, for example.

Tacit meanings don't have syntax in the same sense as natural language, but the system has a certain kind of structure. It consists of structured (schematic, habitual) activity of living organisms in the three-dimensional structure of the physical everyday world. There are individual objects, properties and relations that may involve various habits, that is, meanings. The everyday world is, indeed, interpreted with meanings, but these tacit meanings depend on the structure of the world. Habitual action must accommodate to the objective conditions of action, and these conditions are based on the physical structure of the everyday world. Tacit meanings do not carve up the world, they function in the carved up world. The world is experienced as possibilities of action, and these possibilities are cognized with tacit meanings, habits of action. The experienced world is structured, and this structure is experienced in various practices. In a sense, the experienced world is constituted by these habitual practices. "Through habits formed in intercourse with the world, we also in-habit the world. It becomes a home and the home is part of our every experience" (Dewey LW 10 1987, p. 109).

Language and linguistic concepts do indeed have an impact on how we experience the world, but it is not advisable to take seriously claims and interpretations that are in direct contradiction with the quite obvious condition that the body is the first and most fundamental prerequisite of our existence, experience and cognition, and that bodily action simply must accommodate to objective conditions action and, among other things, prefer doors to windows when exiting a room.

The outcome of this is that yes, the world is always interpreted with meanings, but it does not follow that the structure and order of the experienced world is due to meanings or concepts. Linguistic meanings have an effect on how we interpret the world, but their influence on experience is not so radical as the classical tradition tends to claim.

References

Bennett, M., & Hacker, P. (2003). *Philosophical foundations of neuroscience*. Oxford: Blackwell.

Dewey, J. (MW 9). (1980). *Democracy and education, the middle works 9*. In J. A. Boydston (Ed.). Carbondale and Edwardsville: Southern Illinois University Press.

Dewey, J. (LW 10). (1987). *Art as experience, the later works 10*. In J. A. Boydston (Ed.). Carbondale and Edwardsville: Southern Illinois University Press.

Donald, M. (2002). *A mind so rare. The evolution of human consciousness*. New York: Norton.

Franks, D. (2010). *Neurosociology. The nexus between neuroscience and social psychology*. New York: Springer.

Heidegger, M. (1986). *Sein und Zeit*. Tübingen: Max Niemayer Verlag.

Hildebrand, D. (2003). *Beyond realism and antirealism: John Dewey and the neopragmatists*. Nashville: Vanderbilt University Press.

Johnson, M. (2007). *The meaning of the body. Aesthetics of human understanding*. Chicago: The University of Chicago Press.

Lakoff, G., & Johnson, M. (1999). *Philosophy in the flesh*. New York: Basic Books.

Määttänen, P. (1993). Action and experience. A naturalistic approach to cognition. Annales Academiae Scientiarum Fennicae B 64, Helsinki.

Määttänen, P. (2005). Meaning as use: Peirce and Wittgenstein. In F. Stadler & M. Stöltzner (Eds.), *Time and History, Papers of the 28th International Wittgenstein Symposium* (pp. 171–172). Kirchberg am Wechsel: Austrian Ludwig Wittgenstein Society.

Määttänen, P. (2012). *Taide maailmassa. Pragmatistisen estetiikan lähtökohtia* (Art in the World. An Outline of Pragmatist Aesthetics). Helsinki: Gaudeamus.
Määttänen, P. (2015). Emotionally charged aesthetic experience. In A. Scarinzi (Ed.), *Aesthetics and the embodied mind* (pp. 85–99). Dordrecht: Springer.
Papineau, D. (2006). Naturalist theories of meaning. In E. Lepore & B. C. Smith. (toim.), *The Oxford handbook of philosophy of language*. Oxford: Clarendon Press.
Polanyi, M. (1969). *Knowing and being*. In M. Grene (Ed.) Chigaco: The University of Chigaco Press.
Quine, W. V. O. (1960). *Word and object*. Cambridge: The MIT Press.

Chapter 5
Mind and Interaction

Mind is usually considered to be a property of the brain or, sometimes, a property of the body. This approach is, however, based on the dichotomy of external and internal established by René Descartes. The view that brain processes alone can be considered as cognitive processes is only a modification of Descartes' views. It has been called crypto-Cartesianism (Bennett and Hacker 2003) and Cartesian materialism (Knowles 2013). But Cartesian background assumptions deserve to be rejected.

5.1 From Internalism to Emergence

The idea that mind extends outside the brain and even the body has more recently become under consideration (Clark 1997, Noë 2004, 2009, Rockwell 2005). Andy Clark distinguishes between homuncular, interactive and emergent explanation in cognitive science and points out, quite correctly, that an explanation appealing to interaction "is usually just a sensitive and canny version of homuncular explanation" while the idea in emergent explanation is that interactions "yield types of adaptive behavior not neatly attributable to any specific inner component or system" (Clark 2007, p. 232). The task is to describe in explicit terms what emergence ultimately comes to mean here.

Various versions of enactivism also emphasize the interaction (Varelaet al. 1992). "Enactivism is based on the notion of cognition as emerging out of embodied action. Cognition emerges from processes of perception and action that give rise to recurrent sensorimotor patterns" (Menary 2006, p. 2). Here, again, the task is to explain what emergence actually consists of. The appeal to recurrent sensorimotor patterns comes near to the notion of a habit of action described in the previous chapter, but more can be said about this issue. Daniel Hutto and Erik Myin have a different characterization. Enactivists maintain that our ways of engaging with the world and others

P. Määttänen, *Mind in Action*, Studies in Applied Philosophy, Epistemology and Rational Ethics 18, DOI 10.1007/978-3-319-17623-9_5

"are mindful in the sense of being phenomenally charged and intentionally directed, despite being non-representational and content-free" (Hutto and Myin 2013, p. 13). The problem is to spell out what intentional directedness is without appealing to internal representations and mental contents. Another way to bridge the border between agent and environment is to take "cognitive systems to be dynamical systems, best explained using the tools of dynamical systems theory" (Chemero 2009, p. 25). A most radical claim is that nothing in a cognitive system is a representation (ibid., p. 67). Now if a cognitive system is defined as a system of organism environment interaction, then there are no external representations either (that is, external to the body but internal to the cognitive system). However, the rejection of the notion of internal representation does not exclude an appeal to external representations.

The idea that intentionality can be explained independently of its traditional context, namely independently of the notion of representation, seems to be problematic. Mental states (internal representations) are traditionally considered to be intentional, that is, they are about their objects (referents). The characterization of intentionality as *aboutness* comes from this. Now what would intentionality be independently of this?

The programme of teleosemantics (Millikan 1984) seems to attempt something like this. It tries to explain how the internal states of a physical system can have representational powers "in terms of the biological functions of these states" (Macdonald and Papineau 2006, p. 1). According to Millikan functions are natural purposes (Millikan 2014, p. 64). However, functions and intentions are not necessarily the same thing, so there is still a problem here. Alicia Juarrero's view of dynamical systems comes close to the present attempt to deal with the problem by analysing an agent's control loop through the environment (Juarrero 1999, pp. 195–213). A control loop is precisely what is needed.

The present version of a control loop is realized through action and perception. The notion of habit of action is the proper concept here because it is inherently a teleological notion. The intentionality of representations is derivative of the intentionality of action, but these intentional representations (sign-vehicles) are external to the body. The notion of the intentionality of representations is based on an analysis of external representations that refer to some object. What justifies the transfer of this notion to literally internal states or processes of an organism?

5.2 The Problem of Internal Representations

The conception that cognition is manipulation of internal representations has been a major trend in cognitive science. The idea was adopted from digital computers. They manipulate symbols and seem to do something intelligent, so maybe the brain functions similarly. Human beings are, obviously, manipulating external symbols, but it does not follow that the brain manipulates internal symbols (or representations). This trend continues the Cartesian line of thinking in that cognition is an internal process. So the dichotomy of external and internal is preserved.

Benedict Spinoza was a consistent critic of the Cartesian tradition. He rejected the conception that ideas, being modes of thinking, are images or words (Spinoza 1955, p. 122). Ideas do agree with their objects (ibid., p. 115), but what is this agreement like? According to Spinoza a true idea of a circle, for example, tells how a circle is construed, and this takes place through movement: one end of a line is fixed and the other end moves (Spinoza 1955, p. 35). This gives grounds for the following thought. When a hand draws a circle, some servomechanism in the brain guides its movement. The connection between the circle and the mechanism (the alleged idea of the circle) is realized through movement of the body, not through perception alone. Does it even make sense to ask whether this mechanism is round or not? In other words, the ability to draw circles does not in any way require literally round ideas in the brain. And in Spinoza's thought the movement of the hand belongs to the thinking of circle. Thus an idea as a mode of thinking cannot be located in the brain. Note that Locke's problem concerning the general triangle does not arise in this line of thought. Kant solved Locke's problem with his Copernican revolution. Synthetic activity, construction, mediates the concept of triangle and different triangles on a plane. Concept gives a method of construction, schema, which can be applied in different ways at different times. Put three dots on paper and draw the connecting lines. The relation between one concept (method of construction) and different products of construction is in no way problematic. (Määttänen 1993, pp. 25–29.)

The ability to draw a circle does not require round ideas in the brain. Similarly, the ability to manipulate external symbols does not require any ability to manipulate internal entities in the brain. The ability to control bodily movement in an appropriate way is enough. The brain is an organ that helps us to control overt behaviour, and the manipulation of external symbols is a specific part of human activity in different environments. According to Merlin Donald the brain is really not at all like a digital computer. It is more like "a very large network of extremely fuzzy analogue computers" (Donald 2002, p. 102). But analogue computing does not employ symbols. And finally there is the unanswered question of who does the manipulation. There is no homunculus in the brain. People make programs for computers, and it is sometimes said that nature has programmed the brains. However, it is hard to see nature as a conscious acting agent capable of programming. The pragmatist law of association gives another answer. Biological organisms are active agents who act in nature, and their brain gets "programmed" as a by-product of this (more or less) conscious activity.

One attempt to give an account of cognition as an internal process is to use information theory. Eliasmith (2003) admits that its application requires a theory of meaning. An ordered set of units (letters or patterns of neural impulses) transfers mental content only on the condition that the sender and the receiver have beforehand a common conception of the meanings attached to the sets of units. But what could be a theory of meaning for internal representations? An appeal to those of external representations does not seem to be helpful. For example, ostensive definition cannot be used because it requires that both the referring unit and the referent are public entities. Internal representations are not. The principle that meaning is use cannot be applied because nobody uses one's internal neural states (or processes) in any way analogical to the way we use external words, tools or other instruments. There are sayings like "Use your brain, idiot!" but this is purely metaphorical.

Another problem in applying information theory is related to coding and decoding. Suppose that some feature of the environment "codes" a message into a pattern of neural impulses in some part of the brain, as is the case according to Eliasmith (2003, p. 504). This pattern is then supposed to move along (as it certainly does) and carry information with it. Now we face the problem of the two meanings of the word "information". In one sense it means meaningful content. This how the word is used in colloquial language. But in information theory it means only the order of the units. An ordered set of units conveys meaningful content only on the condition that it becomes decoded. But what or "who" does the decoding. Again, there is no homunculus in the brain. Without an explicit account of meaning, coding and decoding information theory simply cannot be applied.

From the viewpoint of Peirce's semiotic theory (to repeat a point already given) it can be said that all the results of skilful brain imaging gives us only two-place relations between internal neural states (or processes) and, for example, external objects. However, according to Peirce, sign-relation is a three-place relation between a representamen (or sign-vehicle), its object and its interpretant. Sign-relations require necessarily these three elements. A representamen can refer to, represent something, only on the condition that it becomes interpreted to do so. What could count as an interpretant for internal representations? Notorious homunculus is lurking again.

It is apparently an outcome of the Cartesian line of thought that mind is considered to consist of universal units that are intentional, that refer to something. As Richard Rorty puts it, "we have no idea of what a mind is save that it is made of whatever universals are made of" (Rorty 1980, pp. 31–32). This and other ways of defining the marks of the mental have only helped philosophers to insist on an unbridgeable dualism between mind and body (ibid., pp. 35–36).

How to deal with the dualism? Rorty's solution is persons without minds. The problem of the relation between mind and body and the problem of how internal mental units can represent external world is solved by dissolving it. There is no mind stuff, no universal entities capable of mirroring the world. The notion of two ontological realms does not make sense. (Rorty 1980, pp. 125–127.)

Maxwell Bennett and Peter Hacker argue along similar lines. According to them the term "representation" is "a weed in the neuroscientific garden, not a tool—and the sooner it is uprooted the better" (Bennett and Hacker 2003, p. 143; see also Bennett et al. 2007). According to Bennett and Hacker the talk about internal mental entities like qualias as well as all talk about how the brain thinks, decides and so on is based on a mereological fallacy. Mereology is about the logic of wholes and their parts. Mental predicates can be attributed to persons, not to parts of persons like the brain. All this talk about the brain as an active agent is a peculiar form of Cartesianism, namely crypto-Cartesianism. What Descartes said about the soul, is said about the brain.

The argument is based on Ludwig Wittgenstein's idea of language games. Words gain their meaning when they are used in the context of other human practices. Mental predicates are meaningful when people attribute them to other people on the ground of publicly observed behaviour. Brains are not publicly observable behaving organisms, which implies that it does not make sense to attribute mental predicates to brains or their parts. In a similar manner the term 'representation' is meaningful when it is used in the context of human practices. People use pictures, words and other symbols for attending other people's thoughts to the referred things. In the brain there are no such practices, which implies that the talk about internal representations does not make sense.

The obvious possibility to react to the problematic nature of internal representations is to reject the notion. Bennett and Hacker attribute mental predicates only to behaving persons. Rorty rejects the whole notion of mind. There is no mind stuff, no internal universals that could mirror the external world.

There is, however, another option. In Peircean pragmatism universals or real generals are habits of action (Sect. 3.2). And habits just are forms of behaviour. Habits are meanings and beliefs, and thus vehicles of cognition. We can follow Rorty in saying that mind is made of whatever universals are made of. Mind's mode of existence is universals' mode of existence, and the connection between the behavioural criteria of attributing mental predicates and that to which they are attributed is more than obvious. Mind consists of universals, that is, of anticipated habitual behaviour in similar circumstances. Mind is not a property of the brain, not even a property of the body (as Bennett and Hacker correctly observe) but a property of the organism environment interaction. The mode of existence of these habits (real generals) is the anticipated future. Present anticipation is based on past practical experience. Habits are vehicles of cognition. Cognition is ultimately anticipating of action. From this point of view the primary task of the celebrated mirror neurons is not mirroring the behaviour of others (which is what they certainly do), but scanning various possibilities of action (or affordances) in the environment. As Franks points out, the world "is known by the *human actions which it makes possible*" (Franks 2010, p. 87; emphasis in the original).

5.3 The Unit of Analysis: The Loop of Action and Perception

John Dewey emphasized strongly the concrete interaction of organism and its environment. This becomes clear already from his critical attitude towards the concept of reflex arc. This notion is too narrow in that it separates the organism from the environment. Instead he suggested the notion of sensorimotor circuit (Dewey 1975, p. 97). The idea is that the objects of the environment play a role in psychological processes. Strictly speaking one cannot talk about interaction in terms of only the other party of this relation. The correct unit of analysis is the interactive system as a whole. In this sense mind extends outside the body. The objects of the environment belong to "the functional organization of mind" (Määttänen 1993, p. 105).

Interaction proceeds through action and perception, but it is not so simple to distinguish between them. Action and perception take place simultaneously. As pointed out earlier, Peirce made this distinction by saying that in action "our modification of other things is more prominent than their reaction on us" while in perception "their effect on us is overwhelmingly greater than our effect on them" (CP 1.324). Following this we can say that the unit of analysis is the loop of action and perception, where the dominant flow of causal effect goes from the organism to the environment through action and backwards through perception. This loop is realized through causal physical processes, which is why the notion is fully consistent with the soft naturalism adopted in the first chapter.

The loop as a unit of analysis is also fully consistent with the pragmatist definition of experience given above. Empiricism relying on perception (Dewey called it sensual empiricism) maintains that experience is receiving causal effects from the world, which are then interpreted with conceptual resources and other internal conditions. Things in the world cause neural states and processes that can be registered with various forms of neuroimaging, for example. This kind of research hopes to find the neuronal correlates of consciousness. Experience starts by perceiving sensory qualities. Cognition proceeds along the line: perception—reflection—decision—action. Pragmatism widens the concept of experience. It consists of perception and action. Pragmatism puts action in the first place in the different line of cognition: action—obstacle encountered—search for new possibilities of action—reflection—decision—action. This change in the concept of experience is supported by findings in brain research, notably mirror neurons. (Franks 2010, pp. 86–87.) These findings show that the world is experienced as possibilities of action. Perceived things start anticipatory motor mechanisms, which can be executed if a decision to act is made.

Anticipation proceeds with internal associative chains created during past practical experience according to the pragmatist law of association. Anticipation brings to mind past experiences, which were outcomes of past habitual action. If circumstances are similar, then actualisation of the habits in mind can be expected to lead to similar experiences in the future. The formation of these chains requires overt activity where the objective conditions of action are encountered and action becomes accommodated to these conditions. And the cognitive role of these chains is revealed only in the context of the loop of action and perception. The so-called neuronal correlates of consciousness get their cognitive significance only if they are considered as elements within the whole loop of action and perception, as an associated chain of internal states (by virtue of the pragmatist law of association) contributing to anticipation of action.

Bennett and Hacker are right in maintaining that mental predicates cannot be attributed to brains, but they don't go far enough outside the brain. Behaving persons are the one's to which we attribute mental predicates, but behaviour is impossible without the environment, and the correct analysis of behaviour, as interaction requires that the role of environmental objects be taken into consideration. Strictly speaking the mind stuff, habits of action as beliefs and meanings, are modes of interaction. The loop of action and perception is the key to the analysis of mind and consciousness. If the loop is broken, mentality goes out of sight.

5.4 Intentionality as the Mark of the Mental

Intentionality has been a mark of the mental at least since Franz Brentano made the analogy between words and mental states. Mental states are, according to this conception, intentional units that refer to objects just like words. The purpose of this analogy was to define psychology as science distinguished from other sciences investigating human beings like anatomy or physiology. Neural states or processes do not refer. They are not intentional units like mental states or representations. This originally Cartesian conception (according to Descartes ideas are about something, but he did not use the term "intentionality") is still with us in the talk about internal mental representations. The dichotomy of internal and external is preserved as a consequence.

Daniel Dennett's account of intentionality as something inherently internal is an example of this. His conception of cognition is based on an analogy with computer programs. A program as a whole is performing something intelligent, like playing chess. The program is composed of subprograms, these again of smaller part and so on until we have simple logical operations that cannot be conceived as intelligent. He says that complex systems (a program, a person or a person's brain) can be divided into simpler elements: "we can make progress by breaking down the whole wonderful person into subpersons of sorts agentlike systems that have *part* of the prowess of a person, and then these homunculi [sic!] can be broken down further into stil simpler, less personlike agents" (Bennett et al. 2007, p. 88; emphasis in the original). Eventually we get agents that are so stupid that they can be replaced by a machine.

Intentionality can be analysed in the same way. A person is intentional, but subpersonal systems have a diminishing amount of intentionality. When does real intentionality disappear? To this question Dennett has only one answer: Don't ask! On the constructive side he says that maybe we can attribute "hemi-semi-demi-proto-quasi-pseudo-intentionality to the mereological parts of persons" (ibid.). This is not particularly informative, to say the least. Anyway it is clear that Dennett's intentional stance is based on the idea that cognition and intentionality are strictly internal matters.

Dennettian line of thought can be followed from the opposite angle. Suppose I take a pen and calculate on paper that $7 + 5 = 12$. Is the pen intelligent? At earlier time there were mechanical calculators with which people could make complicated calculations quite fast by adjusting numbers on the cover and turning the handle. Is the machine intelligent? Digital computers make extremely many calculations very fast. They are often called intelligent. At which point does intelligence come into picture? What does one mean by intelligence here? This line of thought suggests that digital computers perform automatically and effectively the work that used to be done with pens and paper, like excavator do effectively the job that was previously done with picks and spades. Of course, digital computers have much to do with intelligence, but this is best revealed when they are put in the context of the interaction between human beings and their environment, the loop of perception and action.

In the present pragmatist framework there is no reason to look for intentionality in the brain. Our ability to understand external representations as intentional units, as sign-vehicles referring to things in past, present and especially in the future is based on the notion of habit. Habit of action is a teleological notion. As pointed out earlier, by virtue of habits the (anticipated) future may have an effect on the present but not on the past. This way of putting the point makes it clear that no strange backward causality is required. Past experience is habit formation, and habits make it possible to anticipate future if the circumstances and the acting agent are stable enough. Similar activity in similar circumstances leads to similar experiences.

The agent's ability to interpret objects of perception as intentional units referring to past, present and future experiences is based on habitual action. This follows from the definition that what a thing means is simply what habits it involves. Any object of perception (or a tone of colour or sound, for that matter) may be a sign-vehicle that is interpreted to refer to past, present or future experiences by virtue of habits. This can take place at the level of tacit meanings quite independently of language. Words gain meaning when they are used in the context of the tacit meaningful practices. This means that the intentionality of words and other sign-vehicles is derived from the intentionality of action.

The loop of perception and action is a precondition of the emergence of intentionality. Functional systems are based on causal feedback loops, like for example the thermostat of a radiator. But there is not a bit of intentionality involved. The emergence of intentionality begins if and when the regulatory knot of the feedback loop begins to move in the environment and guides its movements on the ground of the effects it gets from the environment. At some point of evolution this guidance develops so that one may distinguish between action and perception. Recall Peirce's characterization that in perception the world's effect on us is bigger than our effect on the environment, while in action it is the other way round. When different, specialized organs realize action and perception, we have the loop of action and perception.

Genuine intentionality of action emerges at the point where there is more than one possibility of action available and the organism is able to choose between them on the ground of its own internal (but not mental) states. Perception guides action, and this enables habit formation and anticipation of consequences of habitual action. The degree of consciousness increases with the organism's ability to make comparisons between alternative ways of action. The number of these alternatives and the extent to which the anticipation can proceed to the future increase the degree of consciousness. Human consciousness is raised on a different level by virtue of the system of external symbols that can be used for conscious anticipation.

Intentionality is the mark of the mental in the sense that habits are real generals, structured forms of intentional action. They constitute the mental. Habits are actualised as interaction, as the ongoing loop of action and perception. Habits are mental because strictly speaking they only exist in the future. Recall that genuine (potential) generality can only be thought of (Sect. 3.2). Generality resides in the future in the sense that an acting agent can think that she performs an indefinite number of similar habitual acts in the future. However, there is no need to postulate a special mental substance. All acts are realized in the material world. The loop of perception and action with its internal anticipatory mechanisms is realized through causal physical processes. But the mental cannot be reduced to the physical in the sense that theories physics or neuroscience could be enough for explaining behaviour if we could have sufficiently accurate knowledge of neural processes. One important reason is the fact that these theories do not contain teleological concepts. Intentional action requires habitual action within the loop of perception and action. To be a conscious subject is to be conscious about the possibilities of action that the environment affords. The degree and quality of consciousness increases with the number and range of the possibilities the subject is able to consider.

5.5 Mental and Physical Causes of Action

Cognition is realized through causal physical processes. Physical causes are thus always involved. What is their relation to mental causes? Jaegwon Kim traces backwards the physical causes of movement of his left foot from the muscles through neural fibres to the brain. Kim assumes that "we have a pretty good neurophysiological story to tell about how such limb motion occurs" to the effect that "the story ends with some neural event in my central nervous system, presumably the firing of a group of neural fibers somewhere deep in the brain" (Kim 1998, p. 64). Then he begins to discuss the relation between this explanation and the intentional explanation in terms of beliefs and desires.

Kim does not tell where the tracing stops and what could be the criteria of stopping "somewhere deep in the brain". This attempt to locate the causes of action in the brain seems to be a presumption without further justification. Kim rules out the role of external environmental and historical factors because "we expect the causative properties of behavior to be intrinsic and internal" (Kim 1998, p. 37). However, this is simply an erroneous way of putting the question; it is a form of crypto-Cartesianism. Some part of the brain are said to make decisions, although they really are decisions of a behaving person. In a similar way Richard Rorty maintains that habits are bodily states that cause movements in the muscles of organisms (Rorty 1991, p. 93). The only difference between him and Kim is that Rorty talks about bodily states instead of brain states. As we have seen, habits are structured forms of interaction. They are actualised through the loop of action and perception. The relation between physical and mental causes is quite different from this point of view. The loop of action and perception (the mental loop) as the unit of analysis opens the door for environmental and historical factors. Habit formation is based on environmental factors (objective conditions of action) and historical factors, earlier practical experience.

Habits are based on earlier experience, which is a complex matter. We have long evolutionary experience accumulated into our biological structure. Bodily organs, brain structures, limbs and sense organs are kind of crystallized habits enabling us to perceive the qualities we perceive and to guide action, to aim at something or to avoid something. Most of us have the habit of using a door and not the window when exiting a room. This is based on long evolutionary experience about the dangers of high places. Then we have long cultural heritage, which we must adopt to certain extent in order to survive in contemporary social environment. Everyone's own individual experience determines what things are learned and to what extent. All these layers have an effect on what we do and how we do it. This background determines to a great extent what qualities we are able to perceive and what kind of affordances or possibilities of action we can find in different situations.

It does not make sense to trace physical causes of movement to an unknown place "somewhere deep in the brain". Habits are formed on the ground of practical experience, step by step on the top of earlier habits. This holds for biological and cultural evolution as well as for individual growth and development of skills. Different layers of experience, layers of acquired habits, have an effect on habitual behaviour. Most habits function subconsciously. If one wants to trace backwards the physical causes of action, one should trace backwards the physical causes through the loop of perception and action down to the early stages of biological evolution. Conscious decision is only a top layer on all this. We do things without knowing the reasons. "One of the main jobs of consciousness is to weave our lives together in a story that makes sense to us and is consistent with our self-conception" (Franks 2010, pp. 70–71).

As to what counts as causes of action from the viewpoint of the present approach based on the notion of the loop of action and perception, there are three sorts of causes of action: external stimuli (things perceived), internal stimuli (needs) and anticipated future (experiences that are probable outcomes of habitual action). Consider eating. There are three kinds of physical causes involved. First, there is an internal sensation of hunger, need for food. Second, there is an observation of food that functions as an external physical cause of perception. Third, there are internal anticipatory mechanisms that were formed during evolutionary experience. Habit formation based on the pragmatist law of association creates internal associative chains, which enable an organism to anticipate that eating the food leads to the satisfaction of hunger. These three physical causes, two internal and one external, form jointly the physical basis of mental causation, but this can be seen only in the context of the mental loop.

It does not make sense to reduce the concept of hunger to some C-fibre firing. Quine maintains that folk-psychological predicates like beliefs and desires can be "explained away" (Quine 1995, p. 86). Instead of talking about hunger we talk about some C-fibres that are firing. This is quite problematic. If psychological predicates like beliefs and desires are really dropped away (like the predicates 'caloric fluid' and 'phlogiston') we have no behaviour left. If these predicates are replaced by neurological predicates we simply don't understand people's activities as behaviour. We might have descriptions like: "Some C-fibres were firing, and this caused the body to move its organs to the effect that some lumps of molecules were transferred inside the body through a hole." But we don't understand from this that a person was eating because she was hungry. As a matter of fact, there is nothing left to understand. Behaviour is gone, and with it the criteria of attributing mental predicates.

Hunger, food and stomach are conceptually interdependent. It is impossible to understand what hunger is without reference to food and stomach (and certain other parts of the body, of course). The conceptual connection is made explicit in the context of the loop of perception and action (with its evolvement in the past). This loop is the residence of the mental. Mental concepts cannot be compared to notions like phlogiston or caloric fluid that belong only to the history of science. Mental concepts like intentionality, beliefs, and meanings are defined in terms of the loop. As pointed out earlier, drop the loop, drop the mental. But there is no reason to drop the loop (and the mental) by appealing to philosophical naturalism. The loop is fully realized through physical causal processes.

The problem may be posed as follows. "How can the mind exert its causal powers in a world that is fundamentally physical?" (Kim 2005, p. 7). This problem is indeed hard if the mind is considered to be strictly internal and capable of having an effect on neural processes (see Bennett and Hacker 2003, pp. 43–67). If the mind is simply identified with the brain, then we loose all the behavioural criteria for attributing mental predicates to anything. But if the mind is a property of organism environment interaction (the loop of perception and action, or the mental loop), then the problem is formulated differently: How can an organism equipped with mind (that is, as an element in the mental loop) have an effect on the physical world? The answer is already given in the question. By initiating action where muscular effort (with of without external instruments) meets the resistance of the world, and the world is thereby changed.

The relationship between mental causation and physical causation is revealed only in the context of the mental loop. This means that if one takes any part of this loop, be it internal neural processes, the effect of the world in sense organs or overt action where the world is changed, and detaches it from its context, then the physical causation involved remains mere physical causation without any sign of mentality.

5.6 Language and Mind

The notion of habits as tacit (non-linguistic) meanings entails that intentionality and mentality can be attributed to acting agents quite independently of the ability to use human languages. Tacit meanings form the basis on which language evolved.

Rorty, Richard. (1980). *Philosophy and the mirror of nature*. Oxford: Blackwell.
Rorty, R. (1991). *Objectivity, relativism, and truth. philosophical papers,* Vol. 1, Cambridge: Cambridge University Press.
Spinoza, B. (1955). On the improvement of the understanding. (R. H. M. Elwes, Trans.). *The ethics*. New York: Dover Publications.
Varela, F., Thompson, E., & Eleanor, R. (1992). *The embodied mind: Cognitive science and human experience*. Cambridge: The MIT Press.

Chapter 6
Facts and Values in Pragmatism

The dichotomy of facts and values is sometimes held as an established conceptual truth. Values are cut off from facts by the famous Hume's guillotine. Values cannot be derived from facts and do not belong to the facts found in nature. Norms and values form thus a hard problem for naturalism, but only for hard naturalism emphasizing the role of the theories and methods of natural science where there is no room for teleological concepts. Soft naturalism is perfectly consistent with the concept of habit, which is a teleological notion.

David Hume's philosophical presumptions, especially his conception of experience and the object of knowledge, can be questioned. This leads to a different conception of what counts as a fact. The pragmatist notion of fact, in its turn, leads to a different conception of the relation between facts and values. From this point of view facts and values are necessarily intertwined in experience.

6.1 Hume's Guillotine

There are two places in Hume's *Treatise* that are quoted quite often. They are in the same opening, the first in the left page and the second in the right page (Hume 1978, pp. 468–469). In the left page Hume says that morality is not an object of science. It consists not in any matter of fact that can be discovered by the understanding. This is because if, for example, one considers a wilful murder, one cannot find any matter of fact or real existence, which can be called vice. In the next page Hume introduces the principle according to which one cannot derive *ought* from *is*. The famous Hume's guillotine separates sharply values from facts, and the character of values and their mode of existence becomes a serious philosophical problem.

P. Määttänen, *Mind in Action*, Studies in Applied Philosophy, Epistemology and Rational Ethics 18, DOI 10.1007/978-3-319-17623-9_6

Between these celebrated passages Hume says something that is not so often quoted. He says that vice and virtue can be compared to "sounds, colours, heat and cold, which, according to modern philosophy, are not qualities in objects but perceptions in the mind" (Hume 1978, p. 469). The reason for the omitting of these lines is pretty clear. In Hume's philosophical framework it makes sense to say that heat and cold are only "perceptions in the mind". But how well does this fit in with the contemporary definition of heat as molecular movement? The sun is not hot, only our perceptions of sunshine are "hot"?

Hume's conception of experience is limited to sense experience. The object of knowledge consists of the hidden causes of our perceptions that, of course, cannot be perceived in themselves. Hilary Putnam points out that Hume entertains pictorial semantics (Putnam 2004, p. 15). If one is to know that something is a fact, one must literally perceive it. If one cannot perceive causality when looking at billiard balls, then one cannot know that there are causal relations in nature. The same holds for values. The mind is something "internal". It perceives the so-called "external" world through the sense organs and wants to know what it is like (if it even exists in the first place). This philosophical framework is outdated.

Contemporary philosophical naturalism has no such problems in the talk about causal relations. One definition of philosophical naturalism appeals to them in saying that the world is causally closed. It is about time to free also value theory from Humean presumptions.

6.2 Facts and Values of an Acting Agent

In pragmatism the notion of experience is wider than that of sense experience. The world is experienced as possibilities of action, and the object of knowledge is the relation between the present situation and the anticipated future situation that is the outcome of habitual behaviour (or controlled operations). The anticipated future situation is now hidden, but it will be revealed if knowledge is valid and anticipation proves to be correct. On this view also a matter of fact is defined differently. The acting agent belongs to the object of knowledge. She does not only perceive the world but lives and acts in that world, and changes the world in acting there. The two situations (the present and the anticipated future situation) are related to each other by virtue of what the acting agent does. Therefore the notion of a matter of fact also includes all these elements: the two situations and the action (or operations) performed by the agent. A matter of fact concerns what will be experienced as an outcome of some activity initiated in the present situation.

Hume's conception of experience is too narrow, and the same holds for his notion of a matter of fact. The object of knowledge is a relation between two situations mediated by habitual action, and the same holds for the pragmatist definition of fact. Like the object of knowledge, a matter of fact is a relation between two situations mediated by successful action. Facts happen. This holds also for perception. Classical empiricism ignores the active and constructive character of sense perception. To observe that snow is white is to use one's eyes in a complicated operation, which, in normal circumstances, has as an outcome the experience of whiteness. The operation is quick and proceeds mostly subconsciously but is anyway an operation that fits with the pragmatist notion of the object of knowledge and that of a matter of fact.

The pragmatist notion of a matter of fact changes the relation between facts and values. There are always numerous possibilities of action available in every experienced situation. It is impossible to execute all the possible activities at the same time. This implies that an acting agent simply must choose between these possibilities. And to choose is to value. Some possibilities are valued higher than all the others. This is not to say that moral deliberation is involved in all choices to act. On the contrary, the point is that value theory is constructed with a bottom-up strategy. Most choices are done subconsciously, and conscious deliberation is built upon the ongoing flow of action. The important thing to note here is the following. All choices, automatically done subconscious choices and conscious choices based on moral deliberation, involve valuation of the anticipated outcomes of action, which belong, when realized, to matters of fact. Facts and values cannot be separated. They are necessarily (or conceptually) intertwined in experience. Hume's guillotine is in deep rust. It holds only as a logical principle. It is true that if premises do not contain value statements, then the conclusions cannot contain any. But why should we exclude value statements from the premises? The satisfaction of biological needs is an objective fact in nature, and as a positive experience it has (in normal circumstances) a positive value. This value is also an objective fact in nature, or is an objective element in the factual relations of animals' life world, to be more accurate. Animals are hungry and act in order to get satisfied.

Human beings are living organisms that act within nature. An acting agent is necessarily a valuating agent. This brings also values into nature as natural properties. Recall de definition of the object of knowledge given above: $\mathbf{S_1} \rightarrow \mathrm{O} \rightarrow \mathbf{S_2}$. The difference between an observed thing in situation $\mathbf{S_1}$ simply as observed and the same thing as having a value (as something good or bad) resides in the relation of this observed thing to the habitual behaviour of the observer. If behaviour is habitual, it is based on earlier experience of acting in similar circumstances. This enables one to anticipate the consequences of action and compare these potential consequences to the desires and needs of the acting agent. This gives the observed thing some value for the agent. Strictly speaking the values are attributed to anticipated experiences, and the observed thing (the fact in empiricist talk) is only a means for achieving this experience. So strictly speaking it is correct to say that, for example, an apple's value for a hungry animal is not a natural property of the apple as such. What is valued is the apple's property of bringing about a satisfaction of hunger when it is eaten, and this is revealed only in the context of organism environment interaction (just like mental properties). The crucial error of Humean empiricism is the separation of the knowing subject from the object of knowledge. The knowing subject lives and acts within nature and belongs to the object of knowledge. Satisfaction of hunger is an objective phenomenon in nature.

One argument for depriving values from nature is based on the character of natural science. It has long ago rejected Aristotelian teleology and appeals only to causal processes, and there is no backward causation. However, the notion of habit is a teleological one because it entails that anticipated future has an effect on present decisions without introducing backward causality. Anticipation of future satisfaction is based on earlier experiences of satisfaction, and the satisfaction of biological needs is a quite common fact in nature. The loop of action and perception combined with internal anticipatory mechanisms (which get wired when a creature acts in nature) gives teleology that is realized through physical causal processes.

According to the so-called naturalistic fallacy it is a logical error to define good in terms of some natural properties like happiness. For it is always possible to ask of any natural property the question: But is it good? So one comes to ask: Is good good? And this does not make sense, as even the proofreading software points out. The accusation of naturalistic fallacy is, however, based on an aprioristic fallacy, which is the conviction that abstract notions like good (or play) should have a one and only one definition that is based on a priori conceptual analysis and explains why different things are valued to be good. Naturalism rejects any absolutely a priori methods of conceptual analysis (without denying the power of abstract thought). The pragmatist notion of meanings as habitual practices entails that meanings are context dependent. There is no point in searching for exact definitions in areas where only family resemblance can be attained. Family resemblance is a consequence of the pragmatist notion of meaning. Words gain meaning when they are used in the context of other practices, and the word "good" comes to mean different things in the context of different practices.

A further consequence of naturalism is that mind is necessarily embodied, and the body determines a physical point of view (as distinguished from a conceptual point of view). Embodied beings cannot avoid physical viewpoints. This is essential also in value theory. Each human being values things from her own viewpoint. It is always possible to ask of any experience valued as good that is it really good from some other viewpoint or from the others' viewpoint. Actually these questions are just those questions that should be discussed in order to make the society better for all its members instead of searching for exact definition of the good independently of human practices.

Aristotle distinguished between *praxis* and *poiesis* by saying that in *poiesis* the goal of action is separate from the action, while in *praxis* the goal of action is *praxis* itself because the goal is good *praxis* (Aristotle 1962, p. 1140b). It is useless to search for good (exact definition of good or a set of unambiguous rules for attaining good) outside the scope of experience, the sphere of existential practices. For a naturalist there aren't any. A couple of thousands years later it can be added that the historical and cultural diversity leads to a pluralism of viewpoints and conflicts of different value systems, but there is still a prospect for an objective analysis of human values.

6.3 Biotechnical Normativity

The view that values and norms are inherently problematic for naturalism is based on a false dichotomy of nature and culture. As cultural beings we are also biological organisms in nature. Culture is a product of nature. It is a system developed by one animal species. According to Dewey earlier ethical conceptions attempted to get values and norms from "the Moral Mount Sinai or out of the a priori blue" (Dewey 1988, p. 219). In other words, the ultimate source of values is either some religion or philosophical conceptual analysis independently of how the world is and what is our experience of it. Dewey wanted to give value theory a scientific basis and said that biology serves as a bridge (Dewey 1988, p. 247). There is a point in the natural law tradition, which considered that human society is a natural phenomenon and that social norms are natural and thus ultimately similar for all human beings. This view of human society and culture is wrong, but it should not be replaced by an opposite and also erroneous view that separates culture from nature. Some element of normativity can be derived from biology.

The notion of biotechnical normativity (Määttänen 2009, pp. 131–133) is, of course, based on the notion of technical norm. If the end is given, then it determines what kind of measures are adequate. One should choose means that get one to the given end. Biology shows that creatures living in nature have a persistent strive to live their lives to the end of it. And here nobody has any choice. Ones born, a creature must live until it eventually dies. The life may be short or long, but simple observation gives the result that most living creatures tend to continue their lives as long as possible. This is the ultimate end that is given to us by nature. And this end determines to a great extent what to do. We have to breathe, drink, eat, get shelter and so on. This is how things objectively are in nature. This biotechnical normativity also gives us an objective basis for value theory. Preservation of life determines certain means.

Biology gives an objective basis to value theory in the sense that embodied beings are objective material entities in nature. Physical viewpoint is an objective viewpoint. There are views according to which perspective corrupts objectivity (Psillos 2000, p. 722). However, this claim ignores the difference between physical and conceptual point of view. Conceptual and theoretical change is always involved in the progress of science, but no amount of theorizing can change our character as embodied beings using material instruments in science, among other activities. This viewpoint is unavoidable but does not corrupt objectivity in the same sense as commitments to some conceptual viewpoint.

The claim that biology gives an objective basis for value theory does not entail that other values should or could be directly derived from biotechnical normativity. Nature and culture are qualitatively different phenomena, and the system of values is complex and heterogeneous. But one basic value is a direct consequence from what is said above. It is the respect for life.

6.4 Values and Emotions

The view that emotions are inner subjective states is a consequence of the dichotomy of internal and external. Emotions are also experienced as bodily states. David Hume could not perceive values among external facts (as he understood them) and came to the conclusion that valuation, morality, is based on feeling or sentiment (Hume 1978, p. 470). Like in the case of the relation between facts and values, also here the opposite holds: emotions are based on values.

Damasio (1996) has put forth a hypothesis that he calls the somatic marker hypothesis. According to it emotions are signs of values. Negative emotions are associated with things that are related to negative experiences and positive emotions with experiences related to positive experiences. Emotions help us make decisions about what to do. Negative emotions make us avoid situations that seem to be dangerous. Run away and think later. Positive emotions tell us to think closer how to get the possible positive experience. For Damasio emotions are heuristic aids of rational thought.

Damasio's views fit well with the pragmatist approach described in this book. World is experienced as possibilities of action, and these possibilities are valuated on the ground of the anticipated experiences that different courses of action are anticipated to bring about. The difference between Damasio and the present view is that emotions are not just heuristic aids of rational thought. Habits of action are meanings and beliefs, and as such vehicles for rational cognition—rational in the sense of means-ends rationality. Habits are rational in that they help organisms to survive in hazardous environment. This holds especially for tacit (non-linguistic) meanings. Habits are vehicles for anticipating consequences of habitual action. And anticipated future has an effect on the present (but not on the past) precisely by virtue of reminding us what sort of experiences are about to follow. As Dewey puts it, "any object that is overt is charged with possible consequences that are hidden" (LW 1 1981, p. 28). Most of the workings of tacit meanings (habits) are subconscious but the outcome of this subconscious cognition become conscious as emotions as signs of the value of anticipated experiences. Habits as beliefs form a body of tacit knowledge about the environment, about the objective conditions of action.

Damasio's view of rationality is manifest in his example of worker bees, which behave as if they predict which flowers are more likely to have nectar. They appear to form probabilities. Then he asks: "How can bees, with their modest nervous system, produce behavior that is so suggestive of high reason, so seemingly indicative of the use of knowledge, probability theory, and goal-oriented strategy?" (Damasio 1996, p. 186). The explanation to this, according to Damasio, is that the presence of reward can influence the motor system toward a particular behaviour. What seems to be astonishing to Damasio is that the preference system of the bees must be extremely small (ibid., p. 187).

The problem of this view is what Bennett and Hacker call mereological fallacy. It is, allegedly, the brain that decides, but how can so small brain do that? Actually Damasio's description of the bees' learning mechanism is clearly a description about how bees form habits. Bees' behaviour accommodates to objective conditions of action. The relevant features in the environment do not form so complicated system that they would require large and complex brain. Probability theory and reasoning strategy is not the sole and primary source of rationality. Habitual behaviour is rational in its own right (see Kilpinen 2000, pp. 50–79). Rationality is not something to look for in the brain. It is expedient behaviour in certain circumstances and manifests in the system of organism environment interaction, in the loop of perception and action. Also tacit knowledge is knowledge that is acquired through long evolutionary experience.

Accumulated experience produces an emotional attitude expressing the summary of the values of the possible experiences the environment affords. "The attitude is precisely that which was a complete activity once, but is no longer so. The activity of seizing prey or attacking an enemy, a movement having its meaning in itself, is now reduced or aborted; it is an attitude simply" (Dewey 1971, p. 183). These emotional attitudes can be about objects (or, strictly speaking, about the experiences to which habitual action involved leads to), situations or more specifically about single qualities. Dewey uses the German word *Gefühlston* (tone of feeling) to express emotional attitudes that have become thoroughly habitual and hereditary (ibid., p. 188).

In his *Art as Experience* Dewey applies these ideas in his philosophy of art. Paintings are expressive because, among other things, lines and relations of lines "have become subconsciously charged with all the values that result from what they have done in our experience in our every contact with the world about us" (LW 10 1987, p. 107). Paintings as a whole and even single qualities have this emotional property, *Gefühlston*, which explains why a painting is emotionally expressive. This idea applies more generally to any work of art. The subconscious working of tacit meanings explains the emotional power of aesthetic experiences. One interesting point that follows from this analysis concerns music. Music's emotionally expressive capacity is a problem because it is generally taken for granted that music does not denote or refer to anything. This is true only on the condition that the role of subconscious mental processes is ignored. The emotionally expressive power of music (and any work of art, for that matter) is explained precisely because single qualities, their mutual relations and the work of art as a whole do refer, albeit subconsciously, to all the previous experience our species has had during the long biological and cultural evolution (Määttänen 2012, 2015).

6.5 The Layered System of Values

Value theory is constructed with a bottom-up strategy just like the system of meanings. The notion of habit is the connecting link here. Habit gives the pragmatist notion of meaning, and habit as a vehicle of anticipation is also a vehicle of valuation. The bottom-up approach goes from biology to culture, from subconscious to conscious valuation. As pointed out before, there is no such thing as one exact definition that would give the necessary and sufficient conditions of something's being good. This follows already from the pragmatist notion of meaning.

Another important point is that there is no *summum bonum* in the sense of self-sufficient intrinsic values that would have no relation to other things. Values that have no relation to other things obviously have no relation to human practices, and values that have no relation to practices are practically worthless. Also values gain their meaning from their relation to human practices. This entails that the valuation of ends is connected to the valuation of means for attaining these ends. John Dewey expressed this by saying that the desired must be distinguished from the desirable (Dewey LW 4 1984, p. 207). Pleasure and happiness are not valuable as such but as connected with means for attaining these values in a steady, sustained and acceptable way. Dewey's talk about ends-in-view is a way to express this connection.

Ends-in-view are based on valuation of anticipated consequences of action. Anticipation, in its turn, is based on earlier experience. In other words, valuation is based on experience, and experience is complex and layered whole. Long evolutionary experience has equipped us with tacit meanings that mostly function subconsciously. As pointed out above, this working of tacit subconscious meanings may become conscious as emotions. Anyway, most practical skills are based on tacit habits of action. Long cultural evolution has equipped our culture with social structures and skills. The social world with use of language, tools, instruments and so on is already functioning as a social reality into which each new individual must get socialised. During individual growth and development every new human being adopts her own combination of available resources with which she must cope with the physical and cultural environment.

The outcome of all this is a complex and sometimes contradictory system of values. Hopeless relativism does not follow because the cultural diversity of values is based on biotechnical normativity. Also when discussing values we have to distinguish conceptual and physical point of view. As embodied beings, members of the same biological species we share the same physical viewpoint as an objective basis that serves as a starting point in the discussion about the present conditions of life, and about how to improve them.

References

Aristotle. (1962). *The nicomachean ethics*. In D. A. Rees (Ed.), Oxford: Clarendon Press.

Damasio, A. (1996). *Descartes' error. Emotion, reason and the human brain*, London: Papermac.

Dewey, J. (1971). The theory of emotion. In J. A. Boydston (Ed.), *The early works 4* (pp. 152–188). Carbondale and Edwardsville: Southern Illinois University press.

Dewey, J. (1988). Theory of valuation. In J. A. Boydston (Ed.), *The later works 13* (pp. 191–250). Carbondale and Edwardsville: Southern Illinois University Press.

Dewey, J. (LW 1). (1981). *Experience and nature, the later works 1*. In J. A. Boydston (Ed.), Carbondale and Edwardsville: Southern Illinois University press.

Dewey, J. (LW 4). (1984). *The quest for certainty, the later works 4*. In J. A. Boydston (Ed.), Carbondale and Edwardsville: Southern Illinois University press.

Dewey, J. (LW 10). (1987). *Art as experience, the later works 10*. In J. A. Boydston (Ed.), Carbondale and Edwardsville: Southern Illinois University press.

Hume, D. (1978). *A treatise of human nature*. In L. A. Selby-Bigge (Ed.). Oxford: Oxford University Press.

Kilpinen, E. (2000). *The enormous fly-wheel of society. Pragmatism's habitual conception of action and social theory*. Helsinki: Department of Sociology, University of Helsinki (Research Reports No. 235).

Määttänen, P. (2009). *Toiminta ja kokemus. Pragmatistista terveen järjen filosofiaa* (Action and experience. Pragmatic philosophy of common sense). Helsinki: Gaudeamus.

Määttänen, P. (2012). *Taide maailmassa. Pragmatistisen estetiikan lähtökohtia* (Art in the World. An Outline of Pragmatist Aesthetics). Helsinki: Gaudeamus.

Määttänen, P. (2015). Emotionally charged aesthetic experience. In D Alfonsina Scarinzi (Ed.), *Aesthetics and the embodied mind* (pp. 85–99). Berlin: Springer.

Psillos, Stathis. (2000). The Present State of the Scientific Realism Debate. *The British Journal of the Philosophy of Science, 51*, 705–728.

Putnam, Hilary. (2004). *The collapse of the fact/value dichotomy*. Gambridge: Harvard University Press.

Chapter 7
Mind in Action and the Problem of Realism

Realism can roughly be defined as a view according to which we can get objective knowledge about the mind-independent real world, while antirealism maintains that we cannot have access to reality because all knowledge depends on internal conditions like conceptual resources. What this debate comes to concern depends, of course, on what one means by the central concepts like mind-independence. What it ultimately means depends, of course, on the definition of mind. Realism debate seems to be connected to the classical notion of mind as something immaterial or at least as something internal as contrasted to the so-called real world. However, if mind is defined as a property of the mental loop, concrete interaction between organism and environment, then the traditional dichotomy of internal and external must be rejected. The notion of objectivity is also at stake. As we have already seen, physical viewpoint is strictly objective and inescapable. It is thus not a problem for objective knowledge in the same sense as conceptual and theoretical viewpoint. The debate between realism and antirealism makes sense only if there is a reasonable distinction between what is real and what is not so real and why so.

7.1 The Manifest Image and the Scientific Image

Scientific realists sometimes refer to the distinction between the manifest image and the scientific image. One example concerns two kinds of tables referred to as Eddington's tables after Arthur Eddington. A table of the manifest image have perceived properties like solidity, colour, shape and so on. As described in the scientific image, a table is a swarm of elementary particles, which are not coloured, don't have a well-defined shape and so on. This distinction between different images continues the traditional dichotomy of real and apparent.

P. Määttänen, *Mind in Action*, Studies in Applied Philosophy, Epistemology and Rational Ethics 18, DOI 10.1007/978-3-319-17623-9_7

The epistemological problem is how to gain knowledge about the theoretical objects of science that are mind-independent unobservables. As science is the best way to gain knowledge, it is sometimes said that the table of the scientific image is the real object while the table of the manifest image is less real because perceived properties depend on internal conditions.

But is it tenable to insist that these tables are distinct entities and that only one of them is a real object? This claim is based on the assumption that, according to scientific realism, the table of the scientific image is independent of our epistemic access to the world and that the theoretical concepts refer to the real world independently of this epistemic access. Theory and reality are simply put next to each other, and they are supposed to match. One source of this line of thinking is Galileo Galilei who said that the book of nature is written in the language of mathematics. For him it was clear that equations describe the physical reality. Experiments helped to choose the correct ones. Mathematics is the key to the structure of reality. This conviction is still widely held among scientists.

However, the conception that the relation between theory and reality is so simple can be questioned. There has to be some explanation of the fit between a true theory and its object. The pragmatist line of thought is that the fit is operational. Symbols gain meaning as they are used in the context of other practices, like for instance the experimental practices of natural science. These experimental practices are epistemic in character. Access to the world with instruments can be called thick epistemic access (Azzouni 1997).

The manifest image and the scientific image are only two viewpoints to the same world. Bodily organs determine one viewpoint, and material scientific instruments determine the other. Bodily organs and scientific instruments are exactly equally real parts of nature, and their objective properties have an effect on what we can find out with them. The manifest image and the scientific image intersect at the level of the physics of the everyday three-dimensional objects. The distinction between gases, solids and liquids, the world of middle-sized everyday objects, is still scientifically valid. It is not imaginable that some future investigation of microcosm or macrocosm would render the basic categories of everyday physics unscientific, false and nonexistent. Scientific knowledge is an extension of everyday experience. A table as an element in manifest image is the same table that can be described as a swarm of elementary particles. The manifest table is exactly as real as the parts of it.

7.2 Instrumental Phenomenology

Also theoretical concepts gain meaning when they are used in the context of other practices like experimentation. The stand that theoretical concepts refer to the world independently of our epistemic access to it would require a theory of meaning to explain how this is possible. The option that meanings are immaterial entities, which are accessed by reason independently of experience and tell us the properties of things beyond our epistemic access to the world, is not available for a naturalist. Embodied beings are tied to a physical viewpoint to the world, and philosophical analysis is the analysis of how the world is revealed from this viewpoint. This is the basic starting point of phenomenology. But there are different versions of phenomenology depending of what notion of experience is involved. A major trend in phenomenology concentrates on the analysis of sense perception. Also Peirce wrote about phenomenology, but action is included in the pragmatist notion of experience.

Instrumental phenomenology emphasizes the role of material instruments (Ihde 1979, pp. 28–50). The use of instruments changes and widens the scope of experience. Instruments mediate and amplify experience. Their objective properties have an effect on what is the result of our interaction with the world when using these instruments. Instruments are with this respect similar to bodily sense organs. The biological structure of the eyes makes it possible to see colours. The world as experienced by a colour-blind person has no colours. Bodily organs and external instruments determine partly how the world is revealed to us in practical experience. The objective properties of these instruments have an effect on the result of this instrumental interaction. As pointed out, there is an analogy between sense organs and external instruments. That we can say and have evidence that snow is white and quark is green depends on this instrumental access.

The results of using instruments depend on the physical features of the instruments. This can be called the instrumental constitution of the experienced world, as contrasted to conceptual constitution emphasized by Kant and his followers. To see the world as coloured is based on the physical properties of the eyes as bodily sense organs. Also the properties of external instruments have an effect on what is discovered with them. These properties not only select but also transform experience (Ihde 1979, pp. 16–27; Azzouni 2004, p. 383; Baird 2004, p. 115). The whiteness of snow and the greenness of quark are things that are realized within the sphere of instrumental constitution. Operations of inquiry have an effect on the world, and what is observed is the joint result of instruments and the world interacting.

The connection to the world through instruments determines a physical viewpoint (or rather a multitude of viewpoints depending on what instruments are used) that cannot be changed so easily. The viewpoint changes as scientists develop new instruments, but it is not possible to get rid of it in experience. Theoretical speculation is not restricted in the same way and is often very fruitful, but the evidence to support new ideas comes only through experience mediated by experimental devices. Experimental devices are constructed on the ground of different theories, but these instruments as such can be taken to be crystallized knowledge, thing knowledge, as Baird says (Baird 2004). And once constructed they are independent of the theories, and may even function in ways that the designers did not think (as the invention of x-rays, for example, suggests). It is not quite correct to say that perspective corrupts objectivity (Psillos 2000, p. 722). There is a difference between a physical and a conceptual viewpoint. Physical viewpoint is determined by our character of embodied beings using physical instruments, which are as objective and real elements in nature as any other.

Instrumental phenomenology plays a role also in the identification of theoretical objects. Theoretical concepts alone are not enough for having access to theoretical objects independently of our epistemic access to the world (as the proponents of scientific realism seem to maintain). At the age of nanotechnology theoretical objects like elementary particles, atoms and molecules can clearly be manipulated. This means that we have an instrumental access, which enables us to get grip of these objects and do something with them. This instrumental access is epistemic access in the sense that it depends on the character of these instruments.

The instrumentally accessed objects cannot be observed with sense organs. That is why they are called unobservables. Why should we count them as entities outside the scope of our epistemic access to the world? From the viewpoint of instrumental phenomenology this is not the case. They are within the reach of experimental (instrumental) epistemic access. Does it follow that this commits one to antirealism in the sense there is no real world independently of our epistemic access to it?

The answer to this question requires that we distinguish not only between observables and unobservables but also between the objects that we interact with and the ones that are outside the scope of our (present) instrumental access. They might be called *interactionables* and *uninteractionables*. Observables are interactionables because we interact with them with our bodily organs. The unobservables, which are within he scope of instrumental access, are also interactionables. John Shook distinguishes between things that are directly observed and things that are instrumentally observed (Shook 2003, p. 335). The important point is that some instrumentally observed things are unobservables, theoretical objects of science. What are uninteractionables? There is no reason to believe that the universe has revealed all its secrets. The history of science shows that new things are being discovered as science progresses. So quite obviously there is something that is not within the reach of instrumental access. We can be realists with respect to it. The principle of ontological symmetry says simply that we as bodily beings exist exactly as certainly as the rest of the universe. Consistent naturalism maintains that we are products of nature within nature.

The borderline between interactionables (directly or instrumentally observed) and the rest of the universe is moving. It moves as science progresses and develops new instruments for experimentation. The border may be fuzzy in the sense that previous knowledge and theoretical considerations give reason to believe that something exists, but there are very few hints about what it might be. Black energy and black matter are obvious examples. Present theory cannot explain observations without assuming that something of the kind exists. The problem is that we don't know much about them because they don't interact, as they say. Black energy and black matter are uninteractionables. However, the borderline is exact in the sense that at the moment some finds a way to interact with them, they change into interactionables and become elements in the proper object of knowledge: $\mathbf{S_1} \rightarrow \mathrm{O} \rightarrow \mathbf{S_2}$. They move in the sphere of instrumental phenomenology and instrumental constitution. As John Dewey already emphasized, epistemic operations affect and change the world. We can only know the outcome of these operations, and this is not the same as the world before these operations. (Dewey LW 4 1984). But this is not to say that aspects of nature outside the scope of present instrumental access would not exist.

The above considerations do not give any support to possible accusations of antirealism. If mind is defined with help of the mental loop of action and perception, then the perceived world literally is within the mind. The mind is a property of the complex system of our interaction with the world. The unobservable interactionables belong to the real world that we can have knowledge about. If someone has a problem with its being within epistemic access, then there is still the unknowable realm of uninteractionables. We can know *that* it exists on the ground of past scientific progress. We cannot know *what it is like* because knowledge, experimental evidence, brings it into the realm of interactionables. Theoretical hypotheses and extrapolations are, of course, possible, but to really know it is to know how to interact with it.

7.3 Truth and Correspondence

Classical correspondence theory of truth says that a proposition is true if and only if it corresponds to a fact. "Snow is white" is true if and only if snow is white. Or: "p" is true iff p. This is sometimes called T-schema after Alfred Tarski. This conception of truth simply puts propositions and facts side by side without any explanation of how a proposition and a fact are related, there is no explanation of the fit (Lakoff and Johnson 1999, pp. 98–102). This works fine in colloquial language. The problems begin to emerge when this is taken as a theory of truth in the toolbox of philosophical realism, in which one would expect to contain a specification what it actually means. How are propositions related to the mind-independent real world?

Scientific realists sometimes call the classical correspondence theory of truth semantic theory of truth, and take it as a criterion of realism and consider epistemic notions of truth as its rivals (Niiniluoto 1999, pp. 100–105). A realist must be able to use the T-schema. One fails to be a realist if one makes use of some form of epistemic theory of truth. According to semantic truth theory, terms and propositions have meanings by virtue of which they refer to the real world. Meanings are supposed to determine the reference to the world independently or our epistemic access to the world. Because our epistemic access to the world depends on the knowing subject, on theories, concepts and other vehicles of gaining knowledge, our theories are not true about the mind-independent real world if the truth-relation is assumed to hold within our epistemic access. This leads to accusations of antirealism.

One argument against epistemic truth is that the denial of truths about things beyond our epistemic access leads to odd consequences. There are statements that clearly have a truth-value but cannot be verified or falsified because no evidence is available. Bertrand Russell's example is the number of sneezes by Winston Churchill in 1949 (Niiniluoto 1999, p. 104; Niiniluoto 2014, p. 169). However, an appeal to examples like this fails to see the difference between the proper philosophical problem and mere practical limitations. The real question is whether we can have truths about things that are in principle, by definition, beyond our epistemic access. Recall the philosophical sources of the dichotomy of apparent and real. Platonic ideas as real entities in contrast to changing and thus "unreal" empirical world could be achieved, by definition, only by rational thought. The object of knowledge for 1700th century philosophers was the hidden causes of sense perception that, by definition, cannot be perceived. This has not much to do with questions like whether Churchill's personal servant failed to do necessary observations in the past.

Semantic theory of truth appeals to meanings, but what are meanings? What kind of theory of meaning is used here? Unfortunately there are not too many explications of this. It seems to be the case, that the proponents of the semantic truth theory have adopted the structure of experience that stems from the 1700th century, namely the view that meanings are mental entities (as separated from experience). These meanings are able to connect words to the unperceived causes of sense perceptions or scientific observations. And if perception (observation) is the only way to have epistemic access to the world, then it follows that the relation between the elements of the truth relation (correspondence), namely the real world and meaningful propositions, remains a problem. As Immanuel Kant observed, if one adopts the classical view of the structure of experience, then one can only assume that there is a real world (the thing in itself that cannot be known) and its alleged effect on sensibility (the character of which cannot be known). Then there is the alleged correspondence between meanings and the real world. What can we know about this non-epistemic relation? It can be argued that causal access to the world is non-epistemic (Niiniluoto 2014). However, this claim ignores the role of instrumental phenomenology and constitution. Instrumental access is epistemic in the sense that it depends on the character of the knowing subject as an embodied being using various instruments.

It is obviously problematic to explain how we can attain truths about the real world independently of epistemic access. One way to do this is to appeal to scientific progress and say that science approaches truth as an ideal limit. Charles Peirce put it this way. "The opinion which is fated to be ultimately agreed to by all who investigate, is what we mean by the truth, and the object represented in this opinion is the real. That is the way I would explain reality" (CP 5.407). However, this kind of ideal limit keeps our knowledge about the real world quite far in the future, and there is no guarantee that this will ever happen.

7.4 Epistemic Access to the World

Fortunately we don't have to wait for the ideal limit to reach truth and reality. The way out of the problems described in the previous section is to realize that they are not good questions. The problems are solved by dissolving them. Truth about theoretical objects is achieved in the same way as truth about everyday objects. The only difference is that our epistemic access to objects of perception involves the use of bodily sense organs as instruments of investigating the world, but the epistemic access to the theoretical objects of science involves also the use of external instruments, experimental devices. Both kinds of objects are under instrumental constitution, and in this sense they are not independent of us and of our instruments for exploring the world. Bodily sense organs and external instruments have an effect on what is the outcome of our interaction with the world.

The principle of ontological symmetry implies that everyday objects (including human beings as biological organisms) and their parts, theoretical objects, are what they are independently of what we say or think about them with language (natural language or scientific theories). They are what they are although it remains a fact that language as well as scientific concepts and theories have an effect on how we interpret our experience. Conceptual viewpoint may distort our conception about reality, but no amount of theoretical reflection can change the basic features of instrumentally constituted part of the world. To repeat, conceptual viewpoint and physical viewpoint are two different things in spite of the fact that these viewpoints are intertwined in experience.

Experience consists of action and perception (with sense organs and external instruments). This interaction is under instrumental constitution. The problem is how to conceptualize it and form theories about it. Words and other symbols refer to the experienced world because they gain meaning when they are used in the context of other practices. In the case of abstract symbols this reference is not direct, but a connection to the world through other symbols is inevitable if these symbols are to be significant for our knowledge about the world. Fiction (science fiction, philosophical fiction or other) is a different matter.

The first and foremost precondition of human experience and existence is the human body. The body with its sense organs is the first instrument of experiencing the world. This entails that the discussion about the ultimate joints of the world discovered by future physics is idle. The conceptualization and identification of these ultimate constituents is relative, not only to the concepts used, but above all to the instruments used in interacting with these entities. Scientific knowledge is an extension of everyday knowledge in the sense that our character as bodily organisms cannot be ignored. Our sensorimotor experience about middle-sized three-dimensional objects cannot be questioned either by science or philosophy. It would be interesting to know what kind of disembodied scientist would come to tell that the distinction between solids, liquids and gases is unscientific and not true. And there is no such thing as conceptual constitution that would really "carve up the world" differently, to rabbit slices or other. Of course anybody can think so, but that remains philosophical fiction as long as one does not give a tenable explication of what is the concept of concept applied in this context. What are concepts and from where and how do they get the power of really carving up the world contrary to our everyday practical (sensorimotor) experience as embodied beings?

Our access to the world is instrumental. This view can be called instrumentalism, but this requires an important qualification. It has been claimed that John Dewey represents antirealism because Dewey, allegedly, holds that theoretical concepts do not refer to anything because theories are only a tool of prediction (Niiniluoto 1999, pp. 114–115). Theoretical concepts are just instruments for organizing experience. Dewey himself calls his conception of science instrumentalism. However, he means by it something different. Yes, Dewey holds that theories are a tool of prediction and that theoretical concepts are instruments for organizing knowledge, but it does not follow that for Dewey theoretical objects do not exists. His interpretation of Heisenberg's principle of indeterminacy is the following. "Heisenberg's principle compels a recognition of the fact that interaction prevents an accurate measurement of velocity and position for any body, the demonstration centering about the role of the interaction of the observer in determining what actually happens" (Dewey LW 4 1984, p. 161). In other words, the first measurement is an operation, a phase in interaction. How can one interact with something nonexistent? Quite independently of what contemporary scientist think about Dewey's statement, it definitely shows that for Dewey elementary particles do exist.

Dewey probably would agree with the view that all state descriptions of quantum-mechanical systems are relations between the systems and measuring devices in action, but it does not follow that micro-systems isolated from measuring devices would be "naked individuals", bare particulars without properties, sort of "propertyless ghosts" (Niiniluoto 1999, p. 149). They are what they are as uninteractionables, quite independently of what we say or think about them, but the only way to get knowledge about them is to bring them within the reach of experimental operations, which move turns them to interactionables. And what we come to know is the outcome of these operations, that is, outcome of our interaction with them. Entities within and without the scope of instrumental phenomenology are different. This should be no problem, since the theoretical objects of science are objects of experimental operations and measurements.

7.5 Embodied Epistemic Truth

The problems of the classical conception of truth as correspondence are due to the accepted definition of the object of knowledge as the hidden causes of perception, unobservable theoretical objects that are independent of our epistemic access to the world. How are meanings and the world outside our epistemic access related? If no explanation is given, then this story is only a useless fiction according to which meanings separated from experience (what would be their mode of existence?) and entities also separated from experience just are related by some unknown mechanism.

The pragmatist definition of the object of knowledge is different. It is defined as a relation between two situations mediated by habitual action (or controlled operations). This change in the object of knowledge gives the required explication of the fit between propositions and facts (which are also defined differently as a factual relation between two situations, see Chap. 6). Thus we have the scheme $\mathbf{S_1} \rightarrow \mathrm{O} \rightarrow \mathbf{S_2}$.

A proposition presented in situation S_1 is true only if an operation leads to situation S_2 where the proposition is verified. Now we have three kinds of entities to which this principle can be applied: observables, unobservable interactionables and unobservable uninteractionables.

The case of observables is clear. The proposition "snow is white" is true only if operation, namely an act of looking, leads to a situation where someone with normal eyesight sees white snow. Whiteness is a state of affairs within our epistemic access to the world. Snow's whiteness is not independent of the properties of the eyes. There is no whiteness in a color-blind's world. The meaning of the term "white" is the habit of using it in the context of practices of perceiving the world with eyes. An act of looking takes place so fast that is not consciously experienced as an operation, but an operation it surely is. The disposition to perceive snow as white in appropriate circumstances is acquired during the evolution. Recall that bodily organs can be considered as crystallized habits. Habits are also beliefs. The belief that snow is white turns out to be true only if the operation gives as an outcome an experience of whiteness.

The same can be said in the case of unobservable interactionables. To verify the proposition that a quark is green one must use some experimental devises and have the expected outcome, an output of the devices that can be interpreted as an experimental observation of a green quark. The main difference between this case and the case of observables is in the used instruments.

As to the uninteractionables, there is no way to verify propositions about them because there is no interaction with them. One can present theoretical hypotheses about the existence of things (like black matter) as abductive inferences. Observed world behaves oddly, and the existence of some entity would explain this surprising phenomenon. Uninteractionables cannot be elements in the object of knowledge because the object of knowledge is defined as two situation connected with controlled operations. So there cannot be any verified truths about them. To know *that* something exists (more or less likely) must be distinguished from *what* it is. As soon as uninteractionables come within the reach of interaction they change into interactionables about which we can have truths on the ground of some operations of inquiry. The change is an effect of these operations, instrumental phenomenology becomes effective, and the borderline between the world we can know about and the unknown world has moved. However, nothing in this changes the analysis given above.

The fit between propositions and the world is operational. The operational fit gives correspondence between propositions and facts. On the one hand there are meanings, that is, habits as real generals that can only be thought of. Habits are schematically structured plans of action in conscious or subconscious cognition. On the other hand there are the objective conditions of action. William James said that truth happens to an idea. What James called ideas are habits as meanings and beliefs. And that what happens to them is the operational correspondence or fit. Sandra Rosenthal refers to Peirce and James who used the analogy of key and lock in explaining the fit between beliefs and the world: a true belief is a tool that fits like a fitting key opens a lock (Rosenthal 2003, p. 49). A slightly different way of putting this is to say that the schematic structure of a belief (habit) fits with the objective conditions of action and the anticipated outcome of operation is attained.

Correspondence is saved and the T-schema can be applied but the crucial difference between this view and classical semantic truth is that operational correspondence is realized within our epistemic access to the world. A proposition expressed in language is true about observables or unobserved interactionables. That what true propositions are about is always constituted within instrumental phenomenology, which is the outcome of our being embodied creatures in a material world. This is why it is proper to call the present notion of truth as embodied epistemic truth. Truth is a relation between theories and instrumentally constituted real objects.

The outcome that we cannot have true knowledge about the world outside the reach of interaction should not bother anybody. It is not a sign of antirealism. The debate between realism and antirealism is based on the dichotomies of classical philosophy. Once these untenable dichotomies are rejected, we can see that there is only one real world and we live in it. The difference between various possible physical viewpoints is not difference in the sense that some perspective would be "more real" or "less real" in the classical sense, that is, being real in contrast to only apparent or illusory. To accept that viewpoints are inevitable does not lead to unrestricted relativism. Truth is perspectival but not relative (Rosenthal 2003, p. 48). Physical viewpoint is determined by bodily organs and external instruments, and they as real and objective elements in the world as any other.

References

Azzouni, J. (1997). Thick epistemic access: Distinguishing the mathematical from the empirical. *The Journal of Philosophy, 94*, 472–484.

Azzouni, J. (2004). Theory observation and scientific realism. *The British Journal of the Philosophy of Science, 55*, 371–392.

Baird, D. (2004). *Thing knowledge: A philosophy of scientific instruments*. London: University of California Press.

Dewey, J. (LW 4) (1984). The quest for certainty. In J. A. Boydston (ed.), *The later works 4*. Carbondale and Edwardsville: Southern Illinois University press.

Ihde, D. (1979). *Technics and Praxis*. Dordrecht: Reidel.

Lakoff, G., & Johnson, M. (1999). *Philosophy in the flesh*. New York: Basic Books.

Niiniluoto, I. (1999). *Critical scientific realism*. Oxford: Oxford University Press.

Niiniluoto, I. (2014). Scientific realism: Independence, causation, and abduction. In K. R. Westphal (ed.), *Realism, science, and pragmatism* (pp. 159–172). New York: Routledge.

Psillos, S. (2000). The present state of the scientific realism debate. *The British Journal of the Philosophy of Science, 51*, 705–728.

Rosenthal, S. (2003). The pragmatic reconstruction of realism: A pathway for the future. In J. Shook (ed.) *Pragmatic naturalism & realism* (pp. 43–53). Aamherst: Prometheus Books.

Shook, J. (2003). A pragmatically realist philosophy of science. In J. Shook (ed.), *Pragmatic naturalism & realism* (pp. 323–344). Aamherst: Prometheus Books.

Conclusion

The dichotomy of apparent and real, as well as the dichotomy of internal and external dichotomy internal and external have dominated philosophical discussion ever since they were invented. Both dichotomies provided solutions to problems that were considered important in the past. The original motivation for these dichotomies has more or less disappeared, but there is more to it. The invention of both dichotomies is based on a firm confidence on the ontological power of mere thought, namely the capacity to create or annihilate real existence.

The Platonic doctrine of ideas saved the ideal of timeless unchanging truths, but the ideas are simply thought into real existence. They are, by definition, beyond the scope of empirical experience. So the only way to gain knowledge about them is to think about them. Mere thought gives knowledge about entities by its own creation. This line of thought was enforced when René Descartes defined consciousness as everything we are conscious about. He was unaware about the subconscious layers of mind, and the contents of consciousness were supposed to be accessible by mere thought, by introspection. This was a methodological basis for philosophers like John Locke to write extensive studies about the functioning of human understanding. And the ones who assume that meanings, concepts, logical propositions and the like can have objective existence independently of how the material world is and of how we experience it continue along similar lines.

However, the argumentation of Descartes is also based on the confidence on the ontological power of mere thought. The independent existence of consciousness was concluded from the observation that one can doubt the existence of one's body while one cannot doubt the existence of the doubt, the *cogito*. But nothing important follows from this difference. Mere doubt cannot really annihilate the body. The doubt can still be the doubt of an embodied mind, as it in fact is. In order to really find out whether *cogito* can exists independently of the body Descartes should have taken some concrete measures in order to destroy the body but this he did not do, as is quite understandable. What's the hurry? Everyone's body will cease to exist some day, and after that point there is all the time in the universe to consider whether *cogito* exists or not. There is not much evidence that it does.

P. Määttänen, *Mind in Action*, Studies in Applied Philosophy,
Epistemology and Rational Ethics 18, DOI 10.1007/978-3-319-17623-9

The classical conception of the structure of experience as sense perception and the object of knowledge as the hidden causes of perceptions is based on these considerations that are not tenable if one admits that we are one animal species in nature and that all the vehicles of cognition have developed during natural and cultural evolution to embodied creatures who live in nature, who think, perceive and act in the midst of various interactions in nature.

We are not spectators of nature from the outside, and the spectator theory of knowledge is in need of revision. An organism living in nature needs to know what to do in order to achieve its goals. Accordingly, the actions of the knowing subject must be included in the structure of experience. Experience consists of action and perception, with or without external instruments. Hidden causes of perception are not of much help in controlling behavior in the world. Hidden causes of perception as the ultimate object of knowledge must be replaced by the anticipated consequences of action, which entails that the knowing subject belongs to the object of knowledge and changes the world while acting in it.

The key notion in this analysis is that of habit of action. Habits are schematically structured forms of action that are formed when similar behavior is repeated in similar circumstances. During habit formation the structure of action is accommodated to objective conditions of action, and when formed they are beliefs about those conditions of action. The world is experienced as possibilities of action, and habits as beliefs are vehicles of thinking about those possibilities of action. Thinking with habits is anticipation of action. This is in accordance with the object of knowledge as redefined. Habits are also meanings. Any perceived object may be a meaningful sign-vehicle if some habits are involved. Habitual action enables one to anticipate the consequences of action related to the observed sign-vehicle, which thus becomes to mean those consequences. This definition of meaning holds for non-linguistic tacit meanings as well as for linguistic expressions, which gain meaning when they are used in the context of other practices. Linguistic meanings are formed on the basis of tacit meanings, and the notable feature of tacit meanings is non-conventionality. Objective conditions of action and the physical features of bodily beings restrict the possible habits involved and, thus, meanings that can be associated with them.

Habits are not internal to the body or properties of the body. They are rather forms or modes of interaction. To take them as beliefs and meanings is a way to criticize the dichotomy of external and internal, which supports the conception that beliefs and meanings (ideas, mental contents) are literally internal, within the mind or the brain as opposed to the so-called external world. The rejection of this dichotomy leads to a crucial change concerning the unit of analysis. The correct unit is not the brain or even the body but the organism environment interaction, which consists of perception and action. The problem of what is the relation between mental and physical is thereby also changed radically. There is no need to ask how meanings, mental contents, intentionality and the like are related to neural processes. The answers to this kind of questions are notoriously hard to find. These problems are solved by dissolving them. The right question is what is the relation of meanings, beliefs and intentionality to the physical causal processes through which the mental loop of perception and action is realized. The notion of habit of action, as a teleological notion enabling anticipation of the consequences of action, gives the answers.

There are internal processes (internal to the body, that is) namely sensations of bodily needs for air, water, food, injuries of the body and so on. There are internal anticipatory mechanisms in the brain created during habit formation, but there is no need to treat these mechanisms as representations. Two-place relations between these processes and things in the world are not enough for these processes to function as representations. But even these internal processes alone are not enough. The world as an object of action and perception is also involved. Cognition proceeds with the mediation of external meaningful entities.

The loop of perception and action as a unit of analysis gives the possibility to take environmental and historical factors into account in explaining behavior. Actually they are necessary elements in habit formation. Some proponents of hard naturalism exclude them because of the insistence that the causative properties of behavior are intrinsic and internal. But this is due to wrong unit of analysis. It is not a conceptual truth that neural processes are the only possible processes for the physical basis on mentality. There are no such things as literally internal intentionality, mental contents or representations. Behavior is an outcome of a complex system of ongoing interactions, a layered system of habits and dispositions, subconscious reactions to environmental cues and long-term conscious planning of activity on the ground of the anticipation of the consequences of behavior.

The inability to see the correct unit of analysis creates futile problems also in explaining normativity in naturalism. Evolution has no goals, neural processes as such are not normative. However, individual organisms and groups of them do have goals. It is a hardwired goal of living organisms to live their life until it eventually ends, and this biotechnical normativity gives an objective basis for a naturalistic value theory. And evolution would not proceed without this one goal of living beings.

The rejection of the two dichotomies of classical philosophy leads also to a re-evaluation of the problems concerning truth and the debate between realism and antirealism. If mind is defined as a property of organism environment interaction, then the notion of mind-independence is changed accordingly. The mind-independent world becomes to refer to those elements of the universe that are not within the scope interaction with the present arsenal of various instruments. There is no reason to deny the existence of such elements, but they are not and cannot be objects of knowledge, as this notion is also redefined.

The difference between observables and the theoretical objects of science is based on the difference between bodily organs and the experimental devices used in science. These instruments determine a physical viewpoint that cannot be avoided and must be distinguished from various conceptual viewpoints based on concepts and theories. They also involve instrumental phenomenology or the instrumental constitution of the world as experienced. The properties of the instruments have an effect on the world as experience by using them. The outcome of inquiry depends on both sides of this interaction mediated by instruments.

Theoretical concepts gain meaning when they are used in the context of scientific practices and refer to theoretical objects as instrumentally accessed under instrumental constitution. This is epistemic access in the sense that it depends on the properties of the instruments. However, this is not a sign of antirealism because both parties of the interaction are equally real and objective elements in nature. The instrumentally constituted theoretical objects are precisely the objects that experimental science deals with.

Realism is often connected to the classical theory of truth as correspondence. Epistemic theories of truth are considered to belong to the arsenal antirealism. The distinction between the physical and the conceptual viewpoint changes the picture. Propositions expressed in language are true about instrumentally constituted objects, observables or unobservable theoretical objects, and these objects are within our epistemic relation to the world. So truth about them is epistemic. But T-schema ("p" is true only if p) can be used in both cases. And there is an explanation of the fit between propositions and the world. It is operational. To know is to know what to do also in the operations of scientific inquiry.

Bibliography

Chemero, A. (2009). *Radical embodied cognitive science*. Cambridge: MIT Press.

Hume, D. (2001). Dialogues concerning natural religion. In J. Fiezer (ed.), South Bend, IN, Infomotions, Inc., 2001. ProQuest ebrary. Web. November 26 2014.

Kant, I. (KdrV) (1956). Kritik der reinen Vernunft, Werkausgabe III-IV. In W. Weishedel (ed.), Frankfurt am Main: Suhrkamp.

Lorenz, K. (1973). *Die Rückseite des Spiegels: Versuch einer Naturgeschichte menschlichen Erkennens*. München: R. Piper.

Peirce, C. (CP) (1932–1958). Collected papers 1–8. In C. Hartshorne, P. Weiss & A. W. Burks (eds.), Cambridge: Harvard University Press.

Index

P. Määttänen, *Mind in Action*, Studies in Applied Philosophy,
Epistemology and Rational Ethics 18, DOI 10.1007/978-3-319-17623-9

M

N

O

P

Q

R

S

MIX
Papier aus verantwortungsvollen Quellen
Paper from responsible sources
FSC® C105338

If you have any concerns about our products,
you can contact us on
ProductSafety@springernature.com

In case Publisher is established outside the EU,
the EU authorized representative is:
Springer Nature Customer Service Center GmbH
Europaplatz 3, 69115 Heidelberg, Germany

Printed by Libri Plureos GmbH
in Hamburg, Germany